Fig. 1. — See p. 59.

Ages.	Periods.		Groups and Rocks.
Age of Man.	Quaternary.		Vegetable soil. *Alluvial.* — Clay, sand, and gravel. *Drift.* — Clay, gravel, and bowlders.
Age of Mammals.	Tertiary.		*Pliocene.* — Sands, clays, and marls. *Miocene.* — Marls, limestones, &c. *Eocene.* — Clays, sandstones, &c.
Age of Reptiles.	Secondary.	Cretaceous.	*Chalk.* — Flint, greensand, limestones. *Wealden.*
		Oölitic.	*Oölite.* — Shales and limestone. *Lias.* — Shales. *Trias.* — Red sandstones and saliferous marls.
Age of Plants.	Permian.		Magnesian limestones and red sandstones. Shales.
	Carboniferous.		*Coal-Measures.* Alternating sandstones, shales, and coal-beds. *Cavernal limestone.* Conglomerate.
Age of Fishes.	Devonian.		*Old Red Sandstone.* *Chemung.* — Shales. *Hamilton.* — Shales. *Corniferous.*—Limestones. *Oriskany Sandstone.*
Age of Mollusks.	Silurian.		*Lower Helderberg.* *Saliferous.* — Marls. *Niagara.* *Oneida.* *Hudson.* *Trenton Limestone.* *Potsdam Sandstone.*
Age of Radiates	Laurent'n Cambrian.		Crystallized limestones, gneiss, and quartzite.
Age of Minerals.	Metamorphic		Clay-slate, mica-schist, gneiss.
	Granitic		Granite.

Our Planet,

its

PAST AND FUTURE;

or,

Lectures on Geology.

by

WILLIAM DENTON.

FIFTH THOUSAND.

BOSTON:
WILLIAM DENTON, PUBLISHER.
1872.

Geo. C. Rand & Avery,
Stereotypers and Printers,
3 Cornhill, Boston.

PREFACE.

While lecturing upon Geology, which I have done for the last thirteen years, in various parts of the United States and Canada, I have been requested many times to write a volume embodying the substance of my lectures, so that those who had heard them in public might be able to read and study them in private. In accordance with their request, I have written this volume; divesting it, as I do my lectures, as far as possible, of the technicalities by which the science of Geology is so frequently obscured, and presenting it in the same order in which it is presented by Nature herself.

I am indebted to Lyell, Owen, Mantell, Buckland, Ansted, Hall, Dana, and many others; and have freely used their works, as they had done the writings of previous authors.

I am also indebted to Prof. Henry A. Ward of Rochester for copies of engravings contained in his "Catalogue of Casts of Fossils." At vast expense, and by years of persevering labor, directed by rare ability and scientific culture, he has collected many hundreds of rare and unique fossils and casts, to the great advancement of Geology in America.

W. D

Wellesley, June 17, 1868.

CONTENTS.

LECTURE I.

LECTURE II.

LECTURE III.

LECTURE VI.

LECTURES ON GEOLOGY.

LECTURE I.

THIS is a wonderful and beautiful world on whose surface we live. There are mountains whose hoary summits are lost in the clouds that envelop them; grassy valleys lying in beauty at their feet; deep canyons that the sun never visits, through which flow rushing torrents continually; lakes that sleep in the arms of verdant hills; rills that leap with tinkling feet over mossy ledges; and oceans tossing in their resistless might, and grinding to powder the precipices that gird them.

What a multitude and variety of organic existences we behold! — tall pines rearing their graceful heads on the hills; mosses carpeting the damp ground in the vales; eagles soaring above the clouds; humming-birds flitting from flower to flower; deer bounding through the forest-glades; squirrels skipping from bough to bough; whales floating like islands in the ocean; animalcules exploring a drop; and, considering all these, *man*, upright-standing, upward-looking, the fruit of the ages and the brain of the world.

Nor is this all: where the interior of the earth is exposed to our view, we discover beds of clay, — red, blue, white, and yellow; sand, gravel, marl; and more solid beds of shale, slate, sandstone, conglomerate, limestone, marble, trap, and granite. In some of these, we find leaves of trees, shells and fishes, and bones of reptiles, birds, and beasts.

Who is so incurious as not to desire to know the history of all these? Who does not wish to know how and when these mountains were heaved, and they first looked proudly up at the stars above them, and down on the world that lay sleeping at their feet? when the rivers first coursed down their slopes, and commenced the work of carrying them to the ocean? when those canyons were carved, and what carved them? when came into being the grasses and trees, the fishes and beasts? and how myriads of them became embedded in the various rocks in which their remains are found? Who would not learn the history of this great world and its wonders?

Where is the book, you eagerly inquire, in which this history is recorded? There is no volume of man's making in which it may be found; no library could contain the thousandth part of it: but it is inscribed on the rocks around us and beneath our feet; they constitute the volume you wish to read, every word of which is written by Nature's own hand. She has kept a faithful record of all her doings, — kept with wonderful accuracy her diary; and we may read her own statement of facts more wonderful than the fictions of Arabian fancy. All unconsciously, the fiery volcano has traced its turbulent history with a burning pen; the coral and the sea-weed, the fishes, reptiles, birds, and beasts of the olden time have written their life-story in the plastic rock, for us to read. All is recorded in them that has been done to bring the world from the original rude state in which it existed to the condition of life and beauty that crowns it to-day. How glad should we be to have the privilege of reading this wonderful book, and of becoming familiar with its contents!

Men have read and studied this great volume for years; and the science of geology includes what they have discovered in it. This science, dry as some persons deem it, is second in interest to none, and is worthy of the consideration of all.

The *farmer* should study geology; for it treats of what most intimately concerns him. "What has a farmer to do with ologies?" says one. Farmers would be better men, as well as better farmers, if they knew more about the "ologies." "What is this through which your ploughshare is moving from day to day, and from which all your wealth is obtained?" The farmer answers, "Dirt or soil."—"And how do you think it was made?"—"Oh! I don't know: made when the world was made, I suppose." Take a pinch of this soil, place it under a magnifying-glass, and you discover a gravel-bank. Most of the soil is made up, you find, of small stones formed by the wearing-down of solid rocks as hard as the bowlder by the roadside. By the constantly-falling rains, by roaring torrents, by grinding glaciers, by swelling frosts, and blowing winds, the solid rocks have been ground during ages to powder; and the soil is that powder. The soil is therefore, at all times, like the rock of which it was made. Where sandstone has been worn down to make soil, we have sandy soil as the result; where shaly rocks have been thus worn down, we have stiff, clayey soils; and the wearing-down of limestones and shales together, as in Illinois and Kansas, gives us rich, loamy soils abounding with the elements of fertility. So much do the soils resemble the rocks from which they are derived, that in chalky districts the soil is white, red in the red sandstone region, and generally of nearly the same color as the rock from which it was formed.

By consulting a geological map, and thus obtaining a knowledge of the rocks found in certain districts, we may know before we visit them the character of their soils, to what crops they are best adapted, and what districts of country will sustain the largest population.

The farmer is, of course, interested in what lies beneath the soil, especially if he is the owner of his farm. Few farmers dream of the extent and value of their possessions. They know how long their farms are, and just how broad; but how many farmers think that their farms are four thousand miles thick, and all theirs? Immediately beneath the soil may lie just what the hungry surface needs to enrich it: geology frequently points this out. Thus in the southern part of New Jersey, where the soil is much like the sea-beach, little better than shifting sand, marl, or green sand as it is termed, was found underlying this; and, on being laid on the land, the result was a great increase in its productiveness: so that, as Prof. Rogers says, "land, which, previous to the discovery of the green sand, sold for two dollars and a half the acre, was subsequently worth thirty-seven dollars the acre." So much has Geology done for the New-Jersey farmer; and she is prepared to do as much for many more.

Where lie the stores of mineral wealth? — of coal, iron, copper, and more precious metals that have been concealed in the dark recesses of the earth for ages? The farmer and the *miner* are equally interested in the answer to this question. Geology, by teaching us what rocks do *not* contain certain minerals, prevents much useless expenditure of time and money; and, by pointing out the rocks in which they are likely to be found, sometimes enables the miner to dig with confidence, and reap the fruit of his labors.

In Canada, an ignorant farmer complained of a singular green soil upon his farm, from which he could obtain nothing but a few waxy potatoes. A practical man of my acquaintance, with some knowledge of geology, saw this soil, and knew at once the cause of its greenness. Some of that very soil was sold by him for twenty dollars a barrel; and where the soil refused to produce good potatoes is now a famous copper-mine. Many a farmer who never saw deeper into his farm than the bottoms of his ditches is complaining bitterly of his poverty; while within twenty feet of his nose lies what would make him richer than Crœsus.

To the *philosopher*, geology is of incalculable value. No science digs deeper, few soar higher. Do you want a foundation for your philosphy, deep, abiding: here you may find it. For want of it, our so-called philosophies are castles of cards, erected to-day, blown down to-morrow. If history, written by the fallible finger of man, and extending over but two or three thousand years, is so important that men wisely spend a lifetime in its study, how much more important a history of events transpiring during countless ages, written by impartial historians, who have infallibly recorded the facts of the past! More than the telescope to the astronomer, and the microscope to the naturalist, is this science to the philosopher. Mysteries that it once seemed impossible to discover are plain to his illumined vision; and, where he saw through the fog but the dim outline, the landscape now lies before him in the glowing sunshine.

Geology should be studied; if for no other reason, for the happiness that it affords. All knowledge increases our capacity for enjoyment; opens new windows to the soul, through which the sunlight of bliss may beam; and yields to the true-hearted student the purest pleasure.

Wearily and sadly the laborer toils in the quarry, digging rock for the lime-kiln. He goes to his work as the remanded prisoner to his dungeon, without pleasure or hope. The rocks are hard, and very useful when burned into lime,—the sum total of his knowledge with regard to them. But a volume on geology is placed in his hands, and it opens his eyes. He discovers that every rock he handles is written within and without with mysterious characters, whose significance he may learn: he deciphers them day by day, and is delighted with the knowledge which they disclose; every move of his crowbar turns over a new leaf; and he finds himself a happy student in Nature's college, furnished with an excellent library and the best of professors, who do not despise him for his thread-bare coat or "clouted shoon."

Destitute of astronomy, the heavens would be to us a sealed book, the moon no larger than a dinner-plate, and the stars but shining points in the revolving sky. But, with the light that this noble science sheds, we behold the rolling worlds around us, and view with mingled awe and delight the midnight vault that bends over us. And what astronomy does for the heavens, geology does for the earth, and much more. These pebbles we trample under foot are volumes that we may read with profit and pleasure: they are travellers who have wandered over many lands; and they have wonderful stories to tell us when we have learned their language. I make, therefore, no more apology for introducing this science of geology to your notice, assured that it will amply repay you for your most devoted attention.

On examining the earth's surface for the purpose of discovering how it came into its present condition, we soon learn that *water* has been an active operator. Around us lie pebbles, every one of which came into its present rounded form by being rolled in water; for every pebble was, without doubt, once part of a solid rock, which has been broken off, and rolled over and over till it acquired its present shape. Even on the tops of hills we find beds of gravel, sand, and clay. The sand was once quartz or sandstone, and has been worn into its present condition by running rivers, or rolled by the waves of lakes or oceans; while the clay or mud has been formed from shaly or slate-rocks in a similar way. The few exceptions to this I shall treat of in a future lecture.

We see at a glance, then, either that the water has been high as the hills, or the hills have been low as the water. If we dig below these loose beds, we shall find in many places layers of sandstone, shale, conglomerate, and limestone. The sandstone, we shall discover, after some study, was once sand; the shale was once mud or clay; and the conglomerate, or pudding-stone as it is sometimes called, is merely pebbles cemented together. In these beds we find leaves and branches of trees that once floated on the water, and shells that once lived in it. At the bottoms of coal-mines, from one to two thousand feet deep, we may see such rocks and such remains in them; and, in many places where rocks have been lifted up and exposed, such beds have evidently been at one time much deeper than this, — some of them miles in depth.

When we learn these facts, we cannot be much surprised that early geological investigators came to the conclusion that the whole crust of the globe had been deposited as sediment at the bottom of a mighty ocean, at one time covering the whole earth: in other words, they supposed that water had been the sole agent employed in bringing the earth into its present condition. But a more extended and thorough investigation soon showed them that this was a one-sided view of the matter, and that another important agent had also been employed.

In the neighborhood of volcanoes were found beds of ashes, cinders, and hardened lava: no one could dispute that these were made by *fire*. At a distance from all volcanoes, beds similar to these were found, that were also evidently fire-made. In England, where geological investigations were carried on a long time ago, and where there is at this time no volcano, rocks similar to the lavas of Ætna and Vesuvius were found. Between Whitby and Scarborough, on the Yorkshire coast, is what is called a trap-dike, extending in a north-western direction for fifty or sixty miles: it is about thirty feet wide generally; but, in places where the lava had evidently overflowed, I have seen it a hundred yards wide. This dike is a crevice, filled with what was once melted matter, either poured out of some extinct volcano, or forced up from below: for, where it comes in contact with limestone, it is either burnt into lime, or hardened into a substance like marble; where it is in contact with coal, the coal, for several feet from it, is converted into soot, but at a greater distance into anthracite coal, and at fifty feet is unaltered. This rock, then, was made by fire. Rocks similar to it in composition are found in nearly all countries, and, like it, are evidently fire-made.

The Palisades, on the Hudson, opposite New York, are made of this kind of rock. I have seen dikes of it around Lexington, Waltham, and Needham, in Massachusetts. In Nova Scotia, there is a bed of trap (as these ancient lavas are called) one hundred and twenty miles long, and from one to six miles broad, — once, evidently, larger; but much of it has been washed away by the waters of the Bay of Fundy, masses from the undermined cliffs falling continually. West of the Rocky Mountains, thousands of square miles are covered with lava-beds thousands of feet in thickness, appearing, in many places, as if they had been poured out but yesterday.

The more geologists became familiar with the earth's crust, the more apparent to them became the evidence of the operation of fire. They found a large class of rocks which were crystallized, and in this respect differed from recent lavas, but in every other respect resembled them. There were no sedimentary lines in them, indicating the action of water, as we find in the water-made rocks: and they were at length classed with the fire-made rocks; for it appeared that they had slowly cooled at great depths, and thus received their crystalline form. The fire-made rocks were found, as a rule, to underlie the water-made rocks: and at last it became apparent, that, at some time in the immensely-distant past, the whole earth had been in an intensely-heated condition, — in fact, a molten mass; and that rocks of great thickness had been formed from the cool ing of this fiery fluid, before water existed on its surface. You may ask what led geologists to come to such a conclusion as that. I will tell you; for this is of the greatest importance. Once let it be understood that the earth was originally a fiery mass, and we can then learn how naturally it has come into its present condition, and see the reason for a thousand facts that come before us, otherwise dark and mysterious.

First, *the temperature of the earth steadily increases with depth.* If we were to sink a well in Massachusetts to a depth of sixty feet, a thermometer at the bottom, carefully preserved from the surface heat and cold, would show no change of heat throughout the year; for, at that depth, neither the heat of summer nor the cold of winter produces any sensible effect. A Fahrenheit thermometer would indicate there about fifty degrees of heat: but, on going down fifty feet farther, we should find the mercury in the thermometer rise to about fifty-one degrees; fifty feet farther still, fifty-two degrees; and, as we continued to go down, we should find the heat continue to rise at the rate of about one degree for every fifty feet.

Experiments establishing this increase of heat with descent have been made in many parts of the earth, — on the frigid steppes of Siberia, in the mines of temperate England, and within the torrid zone. There is a well at Jakutsk, in Northern Siberia, dug through frozen ground to a depth of three hundred and eighty-two feet. At a depth of fifty feet, the temperature was seventeen degrees, and at the bottom it was twenty-six degrees; being an increase of one degree for about thirty-seven feet: and, at the same rate, it would require a still farther depth of more than two hundred feet to reach ground free from the influence of frost.

At Sunderland, in England, there is a coal-mine, a part of whose workings are fifteen hundred feet below the level of the sea; and there the increase of heat is one degree for every fifty-nine feet.

At the Guanaxato silver-mine, in Mexico, the mean temperature at the surface is sixty-eight degrees; and at the depth of 1,713 feet the temperature is ninety-eight degrees, being an increase of a little more than one degree for every forty-five feet. Humboldt, in his "Cosmos," says, "It is worthy of notice, that, wherever the observations have been conducted with care and under favorable circumstances, the increase of the temperature with the depth has been found to present, for the most part, very closely coinciding results, even at very different localities." Prof. Phillips says, "It appears, then, fully ascertained, that in situations far removed from volcanic action, in different kinds of rock, with very different chemical relations, water, air, and rocks continually grow warmer as we descend in the earth. *Without a single exception*, the interior of the globe is warmer than the surface; and the heat augments constantly with the depth."

In the English coal-mines, at a depth of six or seven hundred feet, the miners find a comfortable summer temperature the year round; so that the hewers, who dig out the coal, work with nothing on but flannel drawers. In a tin-mine near Redruth, Cornwall, at a depth of eighteen hundred feet, there is a permanent heat of one hundred degrees, which is that of one of our very hot summer-days. The miners, it is said, work naked, and ascend, every hour or so, several fathoms, and dip themselves in pools which are comparatively cool, to enable them to labor in this elevated temperature. In the coldest winter-day, then, we are less than half a mile from the warmest summer-weather.

Artesian wells, which are wells bored till the water flows out, instead of being pumped out, reveal in like manner the earth's interior heat. At Grenelle, near Paris, is an artesian well nearly eighteen hundred feet deep. The water flowing from it, at the rate of more than five hundred gallons a minute, has a heat of eighty-two degrees, being an increase of heat from the stratum of invariable temperature of one degree for every fifty-nine feet; being the same rate of increase as in the deep English coal-mine. At Salzworth, in Germany, is an artesian well bored for salt water to a depth of 2,144 feet, from which the water flows at a heat of ninety-one degrees.

A remarkable artesian well has been recently bored at Louisville, Ky., to the depth of 2,086 feet. The water rises in pipes, by its own pressure, one hundred and seventy feet above the surface, and pours out at the rate of two hundred and thirty gallons a minute. As it flows from the top of the well, it has a constant temperature of seventy-six and a half degrees; but, when a registering thermometer was sunk to the bottom, it indicated a heat of eighty-two and a half degrees, or an increase of one degree for about every sixty-five feet. If the heat continues to increase at the same rate with increased depth, as there is every reason to believe, at a depth of a little more than a mile and a half we should find a temperature of two hundred and twelve degrees, or the heat of boiling water. About five miles down, it must be hot enough to melt lead; twenty miles down, hot as melted gold; and, at a depth of forty miles, the heat is more than four thousand degrees; and there are few rocks that could remain solid at such a tremendous heat as this.

The crust of the earth has been variously estimated at from ten to one thousand miles in thickness; but it doubtless differs in different places. The immense pressure to which rocks must be subjected in the interior of the earth may cause them to be solid at greater depths than they otherwise could be; but it is not probable that the crust of the earth is anywhere more than a hundred miles thick, while in many places it may be not more than twenty.

It may be considered certain that we stand upon a rocky crust resting upon a fiery ocean; this crust bearing about the same proportion to the ocean within that the shell of an egg bears to the fluid contents of that egg, or that a coat of ice an inch thick bears to a sixteen-feet-deep lake over which a boy may be skating. Ask a fly what he thinks of an egg, and he says, "It is a mountain of marble." When you hint at the fluid condition of the interior, "It cannot be possible! Have I not walked over it, climbed its mountains, explored its valleys, and stamped upon it with my mighty foot? Solid, solid: it cannot be otherwise." I have heard men talk in the same way, and for the same reason, — they knew no better. Sir John Herschel says, "The central heat of the earth is no scientific dream, no theoretical notion, but a fact established by direct evidence up to a certain point, and standing out from plain facts as a matter of unavoidable conclusion in a hundred ways." Prof. James D. Dana says, "The fact of the existence of the globe at one time in a state of universal fusion is placed beyond reasonable doubt."

We know, from other facts that I shall notice by and by, that the earth has existed for millions of years, during all that time cooling, and its crust thickening: hence we are carried back to a time in its history when its crust was very thin; and a time back of that, when there was no crust upon its fluid surface, but it hung in space a fiery drop.

A second reason why geologists believe in the original fluidity of the earth is its *peculiar shape.* We say the earth is round ; and, as we generally use language, this is correct: but it is not absolutely correct; for the diameter of the earth from pole to pole is about twenty-six and a half miles shorter than from the equator to the equator. Why should it be so? The earth revolves on its axis so rapidly, that a body on its equator is carried more than a thousand miles an hour. It is well known, that, when a body is revolving rapidly, there is a tendency in the matter composing it to fly from the centre, by virtue of what is called centrifugal, or centre-flying, force. We see this force operating when a grindstone is turning quickly. Let water be poured on it, and the drops fly off on every side ; and sometimes large grindstones thus revolving have split to pieces, so strong was the tendency of the particles composing them to fly from the centre. There is the same tendency in the particles of matter composing the earth to fly from its centre as it rapidly revolves. Had the earth been absolutely solid, they could not have obeyed this tendency; but the shape of the earth proves that they have, thus bulging out the earth at the equator: hence they were once free to move, or, in other words, in a fluid condition. This could not have been from the water composing the earth; for, at but a short distance below the surface, it is too hot for water to exist. We cannot avoid the conclusion, then, that the earth was once a molten globe, and its motion spun it into its present shape: for a mathematician can take the mass of the earth, its size, and the rate of its revolution, and calculate what shape its centrifugal force would give it; and the very shape that it should be, according to his calculations, is the identical shape it possesses.

Hot-springs furnish us with another evidence of the internal heat of the globe, and give us reason to believe in its original fluidity. They are widely distributed over the face of the earth. We find them in Greece, Italy, Germany, England, on the slopes of the Alps and Andes, in numerous islands of the Pacific Ocean; while Iceland furnishes boiling-fountains. Nor is the United States destitute of them. In Virginia, Arkansas, Colorado, Utah, and California, they are numerous; and the water in some of them has a heat nearly as high as the boiling-point.

Near the Sahwatch River, in the San-Louis Valley, Col., there is a boiling-spring of immense size, at which hunters sometimes cook their provisions. The sound of its escaping steam can be heard, as I was informed, at a distance of a mile and a half.

In the Middle Park, Col., I examined several hot-springs near Grand River, having a temperature of from ninety-six to a hundred and fourteen degrees: they seem to be situated along the line of a crevice caused by upheaval. At Idaho, ten miles from Central City, in Colorado, several similar hot-springs exist, which appear also to be on a crevice. At this place the ground was so hot, that miners washing for gold were compelled, in consequence of the heat, to stop digging at a depth of twenty feet. Hot-springs are natural artesian wells made by crevices going down to immense depths.

The geysers, or spouting hot-springs of Iceland, are probably connected with the volcanic eruptions to which the island has been subjected; but those of Sonoma County, Cal., may, I think, be attributed to the earth's interior heat. They consist of a number of boiling-springs in a deep ravine, emitting large quantities of steam with a hissing and roaring noise. One of them, called the "Steampipe," sends up a volume of steam more than a hundred feet high.

In the Colorado Desert are boiling-springs which send up fountains of water, accompanied with steam, to a height of twenty or thirty feet. It is said that the escaping steam can be heard at a distance of ten miles.

On one of the Feejee Islands, there is a basin forty feet deep, the water in which has a temperature of two hundred and ten degrees: the natives use it for boiling their food. At Baden, in Germany, there are thirteen warm springs: one of them pours out in twenty-four hours four thousand three hundred and forty cubic feet of water at one hundred and fifty-three degrees of heat.

Sometimes hot and cold springs are found side by side. Thus Homer says, —

> "Next by Scamander's double source they bound,
> Where two famed fountains burst the parted ground:
> This, hot, through scorching clefts is seen to rise,
> With exhalations streaming to the skies;
> That, the green banks in summer's heat o'erflows,
> Like crystal clear, and cool as winter's snows."

One of these springs comparatively shallow, and hence its coolness; the other from some deep-seated source receiving its continual supply of heat.

Hot-springs of Greece, referred to by Herodotus, are still flowing, though we know that more than two thousand years have elasped since they were noticed by the historian. If a woman keeps a kettle boiling for a day, it requires considerable fuel to do it. How, then, does Dame Nature keep her large kettles boiling (some of them forty feet deep, like that on the Feejee Island), thousands of gallons constantly flowing off, and this supplied by cold water to be heated, and that not for a day merely, but for thousands of years? The heat to do this evidently comes from this grand reservoir of heat in the interior of the earth, which, having given up heat to supply the numerous hot-springs of the immense geologic periods, must have been at one time much hotter than now; and we can readily go back to the time when the whole earth was so intensely heated, that water could not rest upon its surface.

If we need any further evidence that the earth was once in a fluid state, *volcanoes* supply it. Humboldt enumerates two hundred and twenty-five active volcanoes; that is, volcanoes that emit vapors at the present time, or have had eruptions within the past hundred and fifty years: and, if all active volcanoes were known, that number would be greatly increased. What are these volcanoes, with their hollow craters, but so many chimneys, communicating with this grand central fire, from which escape smoke, ashes, and devouring lava? They are found from Iceland, in the frozen north, to Victoria Land, in the still more frozen south; from hills a few hundred feet high, to the mighty volcanoes of South America, like Cotopaxi and Antisana, whose blazing cressets, nineteen thousand feet above the sea-level, have been seen on the Pacific Ocean more than a hundred miles from the coast. Commencing at Chili in South America, from there to Equador, volcanoes are found along the Andes over every degree of latitude: passing in a line through Central America and Mexico, and along the Rocky-Mountain chain, they cross by the Aleutian Isles, which seem like so many stepping-stones from the new continent to the old, and contain at least thirty-four active volcanoes, to the peninsula of Kamtchatka; and through Japan, and connecting volcanic vents in the islands of the Pacific Ocean, to New Zealand; forming a grand volcanic chain twenty-six thousand miles in length.

The phenomena of these active volcanoes present the strongest evidence of the present internal fluidity of the earth, and carry us back by fair reasoning to the time when the entire globe was in the same state that its interior now is.

Vesuvius is a name familiar to every one; for who does not remember the first geography possessed at school, and the smoking-mountain that figured in it? Previous to the year 79 A.D., Vesuvius had never been seen in a condition of activity; or, if seen, we have received no account of it. Pliny, the Roman naturalist, does not include Vesuvius in his list of volcanoes; though Strabo, who travelled through Italy before this time, recognized its volcanic character, the lava and cinders surrounding it revealing its true nature to this keen observer, who says it was extinguished for want of fuel. A spectator in the summer of 79 saw, on the coast of the Bay of Naples, two beautiful cities, — Herculaneum and Pompeii, — cities occupied by a remarkably intellectual people. Beautiful villas adorned the neighboring eminences, occupied by the Roman nobility, who came to spend the summer months in what is even now one of the most beautiful spots on the globe. Back of these cities towered Mount Vesuvius, clad with vine-yards and olive-yards, and crowned with a belt of chestnut-trees. A more lovely scene of beauty never smiled on the soul of an artist. The streets of the cities were crowded with pedestrians, the stores full of busy customers, and the temples attended by devout worshippers. In August of that year, the people were aroused one evening to a sense of danger by low, rumbling sounds that were heard beneath their feet, as if the thunder had forsaken the heavens, and taken refuge in the earth; and this followed by numerous earthquake-shocks. The next day, about one o'clock in the afternoon, from the summit of the mountain rose a dark cloud, to which all eyes were turned. It had the shape of a pine-tree when first seen, but gradually extended until it darkened the sky, and a purple twilight settled down upon the land: while from the cloud fell ashes as falls the snow in the winter-time; and out of the dim dwellings poured the distracted multitudes into the narrow, crowded streets. The twilight deepened till a darkness of more than midnight enshrouded the land, — a terrible darkness, only relieved for a moment by flashes from the mountain, that, like lightning, pierced the gloom. The ground rocked, the sea roared, ashes and stones fell in a constant shower; while the terrified inhabitants, some with pillows on their heads, fled for their lives. Down came the ashes for days: foot by foot they rose, till house, temple, steeple, all were covered; and, when men came back to find their old homes, there was nothing to be seen but a desert of cinders, dust, and ashes, everywhere; and no spot previously known could be recognized. Its appearance before and after the eruption is thus described by Martial, a Latin poet, who was living at the time, and had without doubt seen it: —

> "Here verdant vines o'erspread Vesuvius' sides;
> The generous grape here poured her purple tides;
> This Bacchus loved beyond his native scene;
> Here dancing satyrs joyed to trip the green;
> Far more than Sparta this in Venus' grace,
> And great Alcides once renowned the place:
> Now flaming embers spread dire waste around,
> And gods regret that gods can thus confound."

It does not appear that any lava flowed out of Vesuvius at this time, though torrents of mud were poured out: but subsequently lava did flow over the spot where Herculaneum was buried; and, on this account, excavation is much more difficult there than at Pompeii.

There lay the cities, deep buried; and thus they slept for sixteen centuries before their day of resurrection came. New towns were built on the sites of the old ones, the people little dreaming of the cities that lay beneath; till in 1713, a well being dug above Herculaneum, the workmen came down upon the old theatre, where the statues of Hercules and Cleopatra were found. Pompeii was discovered, forty years afterward, by a peasant who was ploughing in his vineyard. Lying nearer the surface than Herculaneum, and only covered with ashes, a large portion has been exposed to daylight, to the great joy of the antiquarian.

The pavement, composed of flags of lava, was found worn by the chariot-wheels, that had been so long still, to a depth of an inch and a half. In the barracks, the scribblings of the soldiers in their idle hours upon the walls were distinctly visible; while in the stocks were found the skeletons of two of them, chained wrist to wrist. Left by their fellows, unable to escape, there they sat till the "fire-shower of ruin" enveloped them.

Paintings were found undimmed by the touch of time; stores just as they were left by their owners, articles lying upon the counter. "I noticed," says M. Simond, "in the Forum, opposite to the Temple of Jupiter, a new altar of white marble, exquisitely beautiful, and apparently just out of the hands of the sculptor. An enclosure was building all round. The mortar, just dashed against the side of the wall, was but half spread out. You saw the long, sliding stroke of the trowel about to return and obliterate its own track; but it never did return; the hand of the workman was suddenly arrested: and, after the lapse of eighteen hundred years, the whole looks so fresh and new, that you would almost swear the mason was only gone to his dinner, and about to come back immediately and smooth the roughness.

"Proculus was a rich citizen; and his house, at the time of the eruption, was undergoing repairs. Painters' pots and workmen's tools were found scattered round. On a bronze dish, a sucking-pig was found ready for the oven, waiting for the bread, seventy loaves of which were found in the oven, turned into something like coal.

"A sentinel stood at the door in his box; the skeleton fingers of one hand holding his weapon, and the other over his mouth.

A woman staid to fill her apron with jewels, but fell in the open court, never to rise again."

The Temple of Isis was well preserved. The skeleton of a priest was found in one of the rooms, and near his remains an axe. He had cut his way through two doors, but was suffocated before the third.

In the eruption that destroyed these cities, it has been calculated that twenty-two million cubic yards of matter were vomited out of Vesuvius. In 1737, there was another eruption, during which there were ejected twelve million cubic yards; and in 1794 another, in which the amount of matter thrown out was estimated at twenty-two million cubic yards. Whence this immense amount of material? The mountain evidently did not furnish it; neither could it come immediately from beneath the mountain, or a cavity would have been formed, into which the weighty mountain would have sunk. It doubtless came from that grand ocean of molten matter, as truly beneath our feet to-day as it was beneath the feet of the inhabitants of Herculaneum and Pompeii.

There have been several recent eruptions of Mount Vesuvius. One, which took place in 1855, is thus described by a correspondent of "The London Athenæum:" "The lava was pent within the deep banks of a wide bed, and was flowing down, not like a fluid, which is the ordinary motion of it, but like a mountain of coke, or, at times, like highly gaseous coal. It split and crackled and sparkled and smoked, and flamed up, and ever moved on in one vast compact body. Pieces detaching themselves rolled down, leaving behind a glare so fierce, that I could have imagined myself at the mouth of an iron furnace; and as every mass fell down with the noise of thunder, or rolled sideways from the upper surface into the gardens and vineyards, the trees flamed up, and the crowds uttered shouts of admiration and regret. Following the course of the stream, or rather tracing it back to its source, we walked by the side of that huge leviathan through highly-cultivated grounds, now trodden under the feet of multitudes, until we arrived at the edge of a precipice; whence we looked into the boiling flood, fed by the cascade of lava which was pouring down from above. Full one thousand feet fell that glowing, flaming Niagara, in one unbroken sheet, over the precipice. Forming, at first, two cascades, the interval between had been filled up by the immense masses of scoriæ which the mountain had thrown out; and now it majestically rolled down, one continued stream, into a lake of boiling fire. There were times when projections in the face of the lava seemed to impede its course; then, behind those projections, accumulated tons upon tons of material. It was a moment of breathless expectation: all eyes were fixed upon that one blackened spot. There was a slight movement; we heard a click; a few ashes and stones fell, and down went a mountain of solid fire into the boiling, smoking abyss, with the noise of thunder."

At this time, Vesuvius is still busy, and the people in its vicinity are watching the progress of an eruption with intense anxiety.

Ætna, nearly eleven thousand feet high, is one of those unruly volcanoes that are never at rest. Virgil says, —

"By turns, a pitchy cloud she rolls on high;
By turns, hot embers from her entrails fly,
And flakes of mounting flames that lick the sky.
Oft from her bowels massive rocks are thrown,
And, shivered by the force, come piece-meal down.
Oft liquid lakes of burning sulphur flow,
Fed from the fiery springs that boil below."

This mountain, ninety miles in circumference, composed as it is of beds of lava and layers of ashes, is a grand monument of the igneous activity of the earth, and bears upon its sides inscriptions which tell of the fluid condition of its interior.

In 1669, a lava current ran from this mountain, and, after overwhelming fourteen towns and villages, arrived at the city of Catania. The people of the city, preparing for such an eruption, had built walls around it, for their protection, sixty feet high. On came the rocky torrent, moving less than a mile a day, but with irresistible force; and, reaching the wall, it rose foot by foot to the top of the rampart, and then poured into the city, ran through it, and beyond it into the sea, in a stream six hundred yards broad, and forty feet deep. The wall of the city is still standing; and the lava stream, hardened, of course, into solid rock, may now be seen curling over it. The lava of this eruption covered eighty-four square miles.

Lyell says, "This great current performed the first thirteen miles of its course in twenty days, or at the rate of one hundred and sixty-two feet per hour, but required twenty-three days for the last two miles; and we learn from Dolomieu that the stream moved, during a part of its course, at the rate of fifteen hundred feet an hour, and in others it took several days to cover a few yards. While moving on, its surface was, in general, a mass of solid rock; and its mode of advancing, as is usual with lava streams, was by the occasional fissuring of the solid walls. A gentleman of Catania, named Pappalardo, desiring to secure the city from the approach of the threatening torrent, went out with a party of fifty men, whom he had dressed in skins to protect them from the heat, and armed with iron crows and hooks. They broke open one of the solid walls which flanked the current near Belpasso, and immediately forth issued a rivulet of melted matter, which took the direction of Paternó; but the inhabitants of that town, being alarmed for their safety, took up arms, and put a stop to farther operations."

In Iceland, there are a number of remarkable volcanoes; none more so than Skapta Jokul. In June, 1783, a torrent of lava flowed from this mountain with a loud, crashing noise; filling up in its passage a deep ravine which the River Skapta had made, which was, in places, two hundred feet broad and six hundred feet deep, and then the bed of a deep lake. One week after this, out burst another fiery river, rolling rapidly over the surface of the first, then farther, damming up various streams, and, after flowing for several days, filled a deep abyss which a tremendous cataract had been hollowing for ages, and again continued its course; nor did it cease to flow till forty miles from its starting-place. This stream was in some places seven miles broad. A few weeks afterward, another stream flowed in an opposite direction for fifty miles; its greatest breadth being from twelve to fifteen miles. It has been calculated that the matter poured out of this mountain in two months would make a solid globe six miles in diameter. Thirteen hundred human beings lost their lives by this eruption, which also destroyed twenty thousand horses, seven thousand horned cattle, and a hundred and thirty thousand sheep. Iceland has not yet recovered from its terrible effects.

The most remarkable of recent eruptions is that of Tomboro, a volcano on Sumbawa, one of the Molucca

Islands, in April, 1815. It is thus described in the history of Java by Sir Stamford Raffles: —

"This eruption extended perceptible evidence of its existence over the whole of the Molucca Islands, over Java, a considerable portion of Celebes, Sumatra, and Borneo, to a circumference of a thousand statute miles from its centre, by tremulous motions and the report of explosions; while within the range of its more immediate activity, embracing a space of three hundred miles around, it produced the most astonishing effects, and excited the most alarming apprehensions. In Java, at the distance of three hundred miles, it seemed to be awfully present. The sky was overcast at noonday with clouds of ashes; the sun was enveloped in an atmosphere whose palpable density he was unable to penetrate; showers of ashes covered the houses, the streets, and the fields, to the depth of several inches; and, amid this darkness, explosions were heard at inter vals like the report of artillery or the noise of distant thunder."

At Bima, forty miles off, the roofs of houses were crushed by the weight of ashes that fell on them; many of them being rendered uninhabitable. The sound of the explosions was heard nine hundred and seventy miles off in one direction, and seven hundred and twenty miles in the opposite direction. Out of a population of twelve thousand, in the province of Tomboro, only twenty-six escaped with their lives. It has been calculated that the ashes which fell on this occasion would have covered the whole of the States of Maryland and Delaware to the depth of two feet, or they would have made a mountain twice the size of Mont Blanc.

On Hawaii, one of the Sandwich Islands, is an active

volcano, Kilauea. Within its crater is a lake of lava thirteen hundred feet below the summit. In 1840, this lake, fifteen miles in circumference, became a vast boiling caldron; the lava rising, like water in a kettle, till it at length burst through the side of the crater, and ran for forty miles, partly underground, and partly above ground; and then poured into the sea, for three weeks, "in a blazing cataract as wide as Niagara," heating the water of the ocean for twenty miles along the coast, and making hills of scoria and sand from two to three hundred feet high. Forty miles off, the finest print could be read at midnight; and the light of the volcano was like that of the rising sun.

Mauna Loa is a volcano on the same island, thirteen thousand feet high. In 1859, a terrific eruption took place, described by various observers. On the evening of Saturday, Jan. 22, the snow on the mountain was seen white, and there were no signs of an eruption; but on Sunday thick clouds of smoke gathered about the mountain, and at night the whole sky was lit up with a terrific glare, and the lava could be seen spouting in a jet nearly one thousand feet high, in the form of an immense pyramid, at times diverging and falling in all manner of shapes, like a beautiful fountain, into a crater about one thousand feet in diameter. After flowing eight days, a stream of lava reached the sea, where it spread out to about half a mile in width, forming in its course several fiery cascades (one of which was from eighty to a hundred feet high), over which the lava poured in a stream moving at the rate of ten miles an hour.

Mr. Coan, a missionary at the island, thus describes another eruption of Mauna Loa as recently as 1866:

"For twenty days and nights, it sent up a fiery jet one hundred feet in diameter, varying in height from one hundred to a thousand feet. As it issued from the awful orifice, it was at a white-heat; as it ascended higher and higher, it reddened like fresh blood, deepening its color, until, in its descent, much of it assumed the appearance of clotted gore. From this fountain, a river of fire went rushing and leaping down the mountain with amazing velocity, filling up basins and ravines, dashing over precipices and exploding rocks, until it reached the forests at the base of the mountain, where it burned its fiery way, consuming the jungle, evaporating the water of the streams and pools, cutting down the trees, and sending up clouds of smoke and steam, and murky columns of fleecy wreaths, to heaven.

"All Eastern Hawaii was a sheen of light, and our night was turned into day. So great was our illumination, that one could read without a lamp; and labor and travel and recreation went on as in the daytime. Mariners at sea saw the light at two hundred miles' distance. The rivers of fire from the fountain flowed about thirty-five miles."

What immense force must be exercised to drive the lava from the interior ocean, perhaps fifty miles deep, to the top of Mauna Loa, fourteen thousand feet above the sea-level! The distance to which large stones are sometimes thrown impresses us likewise with the exercise of prodigious force. Cotopaxi threw a stone, one hundred and nine cubic yards in volume, nine miles.

These eruptions, occurring in such widely-separated portions of the globe, and so frequently, are strong evidences of the fluid character of its interior.

Extinct or dead volcanoes tell the same story, and

point more emphatically to the original fiery condition of the whole. If the earth was once a fiery fluid mass, we may naturally suppose that volcanoes were once larger and more numerous than at present; and this is just what we find. The Peak of Teneriffe, twelve thousand feet high, stands in the centre of a volcanic plain containing a hundred square miles, surrounded by perpendicular precipices and mountains, which were once the border of the ancient crater. This plain was a lake of boiling rock, dashing its fiery waves against those precipitous rocks by which it is now bounded.

In a space twenty miles long and ten broad, between Naples and Cumea, in Italy, there are no less than sixty craters, some of them larger than Vesuvius. One of them is two miles in diameter. The city of Cumea has stood in the centre of one for three thousand years. Vesuvius itself is merely a small cone erected within the crater of a large volcano.

In France, extinct volcanoes and their products cover thousands of square miles. They exist in Spain and Portugal, Germany, Asia Minor, Mexico, and the Rocky-Mountain region. There are hundreds in the Island of New Zealand alone. On Hawaii there is an ancient crater fifteen miles in circumference, and another twenty; while in Maui, a neighboring island, there are two others, twenty-four and twenty-seven miles in circuit. From these facts, we are naturally led back to the time when the earth was one grand volcano, with no crater to confine its fiery waves.

Volcanoes are not even confined to the land. In 1831, a volcano burst up from the bottom of the Mediterranean, off the coast of Sicily, and was called Graham's Island. One month after its appearance, it was one hundred and eighty feet high, and one and one-third miles in circumference. It disappeared, however, in a few years, leaving only a rocky shoal.

In 1811, there was an eruption of cinders at the bottom of the sea near St. Michael, one of the Azores. In six days, an island was formed three hundred and twenty feet high. It was the third time that an island had appeared and disappeared near the same spot.

In the neighborhood of the Aleutian Isles, in 1796, an island arose from the sea, which continued burning for nearly eight years. In 1819, this island was sixteen miles in circumference, and more than two thousand feet high. As I write this, we learn from "The London Times" that a new Greek island has just risen from the bottom of the Bay of Santorin. The bay is about six miles long and four broad, and contains three islands that have risen from the sea during the historical period. On the 31st of January, a noise was heard like the firing of artillery; on the next day, flames issued from the sea to the height of fifteen feet; on the 4th of February, the eruptions were more violent, gas being forced up from the depths with terrific noise; next morning, the new island was visible, increasing sensibly to the eye, till it attained the height of a hundred and thirty feet, a length of three hundred and fifty, and a breadth of a hundred feet. The temperature of the sea in the vicinity rose from sixty-two degrees to a hundred and twenty-two degrees; which is hotter than the hand can bear.

The cooling influence of water for untold ages has not been able to quench the internal fires of the globe; and we are ever receiving evidences of what must have been the condition of our planet when heated to the surface over its whole extent, and no water could possibly exist.

Earthquakes are strong evidences of the present fluid condition of the earth's interior. If the earth is solid to the centre, how can it "shiver and shake, and quiver and quake," and rise and fall like a boat on a stormy sea?

At Lima, in Peru, there are on an average forty-five earthquake-shocks in a year. In Europe and adjacent parts, observations show that there are on an average about forty earthquakes a year; and it is probable, that, taking the whole globe into account, we have on an average an earthquake a day.

In 1692, the Island of Jamaica was visited by an earthquake, which caused the ground "to swell and heave like a rolling sea." In thousands of places, the earth opened; and plantations, villages, and cities sank to rise no more. A tract of land of a thousand acres around Port Royal, the capital, sank down in less than a minute during the first shock; and the sea immediately rolled in. A frigate that was repairing at the wharf was driven over the tops of many buildings, and then thrown upon one of the roofs, through which it broke. Mountains were shattered; the courses of rivers changed; new lakes were formed, and old ones obliterated.

In 1812, the Valley of the Mississippi was terribly shaken from New Madrid to the mouth of the Ohio River. The earth was raised in waves, so that the tree-tops stooped to the ground, and, as the wave rolled on, recovered their erect position. Lakes were formed several miles in extent, and chasms opened in the woods, so that in some cases men chopped down trees, and sat upon them, to save themselves from being swallowed. When these convulsions ceased in the Mississippi Valley, La Guayra and Caraccas, cities in South America, two thousand miles distant, were destroyed. The violent earthquake of Guadaloupe in 1842, which destroyed several towns in the West-India Islands, extended from Charleston, S.C., to the mouth of the Amazon River. The earthquake-shocks felt in Norway are always connected with volcanic disturbances in Iceland.

Humboldt states "that in 1797 the volcano of Pasto, east of the Guaytara River, emitted uninterruptedly for three months a lofty column of smoke; which column disappeared at the instant, when, at a distance of two hundred and eighty miles, the great earthquake of Riobamba, and an immense eruption of mud, took place, causing the death of between thirty thousand and forty thousand persons." On the night in which Lima and Callao were destroyed by an earthquake, four new volcanoes broke out in the Andes.

There appears to be a subterranean communication between Vesuvius and Ætna; so that, almost invariably, when one is active, the other is quiet. The communication thus found to exist between distant volcanic vents and earthquakes at widely-separated places points to the great interior ocean with which they are connected.

A very remarkable earthquake occurred at Lisbon in 1755. It was the 1st of November, — All Saints' Day. The churches were crowded with worshippers, when, about nine o'clock in the morning, they were startled by that subterranean thunder so generally accompanying earthquakes; and in a moment the walls crashed, the steeples toppled, and, in six minutes, sixty thousand people had lost their lives, — thirty thousand killed by falling churches alone: for an earth-wave passed under the city, at the rate, it is said, of twenty miles a minute, though accuracy cannot be expected in calculations made at such times; and its effects were felt for immense distances. The largest mountains of Portugal were split and shaken from their very foundations. Lead-miners in Derbyshire, England, heard the grinding rocks, and hurried out of the mines, fearing to be enclosed in a stony prison from which there could be no release. The waters of a hot-spring at Bath suddenly became red, and hotter than usual. In the Islands of Antigua and Barbadoes, in the West Indies, where the tide usually rises little more than two feet, it suddenly rose above twenty feet; and the water became black as ink. In fact, the whole bed of the Atlantic seems to have been raised. The waters of Lake Ontario were observed to be unusually agitated. At Algiers, in the north of Africa, the earth was as much shaken as in Portugal; and a town in its neighborhood, of eight thousand inhabitants, was swallowed. "It has been computed," says Humboldt, "that a portion of the earth's surface four times greater than that of Europe was simultaneously shaken." Others have computed that this convulsion sensibly affected one-twelfth of the area of the globe. How deep-seated must have been the cause that could produce so wide-spread an effect!

What can that cause be? Taking the wing of a dusky demon, descend with me into the nether regions, and let us, if possible, discover what it is that shakes the earth so disastrously. Through soil, sand, and gravel, and alternating strata of sandstone and limestone, — the graveyards of unnumbered generations, — we descend for thousands of feet. Here is a mineral vein: by it we can descend more readily. What masses of lead, in sheets and gigantic cubes, lining immense caves! But we cannot remain long to admire it. Down again, through mica-schist, gneiss, granite, and quartz, in beds of immense thickness; on through rocks white as the driven snow, and sparkling as that snow in the sunlight of a frosty morning. What is that shining substance? How bright and beautiful! It is gold, — a lake of solid gold: once fluid, its surging waves went sweeping along its brilliant surface, but now still as a "painted ocean." It has sunk in the centre as it cooled; and there is a grand golden amphitheatre left, a mile in diameter. Down again, resolutely, through the centre of this golden floor. Now rocks fly past us, inky-black. We are diving through the ribs of the earth to its warm heart; and a night dark, hot, and stifling, closes around us. In the distance we behold a faint gleam. The rocks are cherry-red, now full red; and now they glow like the walls of an iron furnace. One plunge, and we stand on the shore of the fiery ocean, and, breathless with terror, look upon a scene such as mortal eye never before beheld. A stillness like that of death hangs over it: and yet it does move; it ebbs and flows like the ocean. It is not always thus calm. At times, there are dreadful storms, when these fiery waves roll and dash in fury; and *a storm on this ocean below is an earthquake on the surface above.*

But what could make a storm on this ocean beneath? What makes a storm on the sea? The sun rarefies the air over one portion of the globe, and thus makes a partial vacuum: the surrounding air rushes in to fill it, and, passing over the water, heaps it up, and rolls it along in waves. I can imagine an analogous cause at work in the interior of the earth. Motion and heat may be constantly transformed into electricity within the shell of the earth more rapidly than it can be conducted through the rocks to the surface. When a sufficient amount has thus been generated, it bursts through in some weak place like Ætna or Vesuvius, producing volcanic eruptions. Thus a comparative electric vacuum is produced beneath these volcanic vents; and the electricity in other portions of the interior ocean propels the fluid mass in waves toward these places, these waves lifting the rock above them to the surface; and thus an earth-wave passes along that surface, corresponding with the wave beneath, and producing most terrible effects.

Some attribute volcanic eruptions to the contraction of the earth's crust, caused by its cooling. Thus Lieut. Portlock says, "The solid crust continues to contract as its temperature diminishes in a greater ratio than the central mass; and, as the velocity of rotation must increase as the diameter of the planet diminishes, there will be a tendency to diverge farther from the spherical form, and hence the fluid matter within will press against the contracting crust, and produce volcanic eruptions. M. Cordier has calculated that a contraction of $\frac{1}{12350}$ of an inch in the mean radius of the earth would be sufficient to force out the matter of a volcanic eruption."

Earth-shakings, and some destructive ones, may have been produced in this manner. Between the fiery ocean and the rocky shell that encloses it, there probably are in places vast arched spaces, against whose sides the lava waves beat at times, and deposit rock. This arch giving way by the contraction of the earth's crust, overhanging masses, weighing millions of tons, drop into the ocean beneath, and produce a wave, which, rolling to where roof and ocean are in direct contact, uplifts the rocks to the surface; and we have, consequently, an earthquake-wave, most violent nearest the place of disturbance, and dying away as the waves of the underground ocean sink into a calm.

Small earthquake-shocks are often produced by masses of rock falling into subterranean cavities: some of these may be heard and felt for great distances. The motion of a passing locomotive can be distinctly felt in some houses a mile from the railroad.

It has been proved by long and careful observation of earthquake-shocks, that the number near new and full moon exceeds the number at the quarters, nearly in the proportion of six to five; and their frequency increases when the moon is nearest to the earth: that is, when the tides of the ocean are highest, earthquakes are most numerous; the influence of the sun and moon upon the internal ocean being still felt and indicated.

I have dwelt longer on this portion of my subject than some might deem necessary, because there are many facts presented to us in the study of geology that can only be accounted for by a knowledge of the original igneous or fiery condition of the earth, and its gradual cooling during long ages. This explains why the lowest rocks are fire-made, why they contain no remains of animals or vegetables, why the climate of Great Britain was once as hot as that of Cuba, and why the remains of tropical trees are found in Central Vermont; how corals existed as far north as Baffin's Bay, and why the earliest rocks formed by water have been changed into crystalline rocks: in short, destitute of this, we have no certain geological guide, and facts are continually presenting themselves that we have no means of explaining.

But what could have produced this intensely heated condition of the earth? It is not positively known; and yet astronomers and geologists are generally agreed as to the way in which they think it was produced. Facts discovered by astronomers during long periods of observation have led to the formation of a theory which accounts most beautifully for the original fiery condition of the earth.

Most persons are sufficiently acquainted with astronomy to know that the earth belongs to what is called the solar system. In this system, the sun is the grand centre, around which a number of bodies are revolving, known by the name of planets. First, Mercury, or, as far as we positively know, the first, at a distance of, in round numbers, thirty-five millions of miles from the sun; thirty-one millions of miles farther into space is Venus; and twenty-six millions farther, our own planet. First outside of the earth's orbit is Mars; then a number of small planets, of which we know about ninety, and perhaps, if we knew them all, we might count more than ninety thousand; then Jupiter, Saturn, Uranus, and Neptune, believed to be the outermost planet of the solar system, at the enormous distance of nearly three thousand millions of miles from the sun.

There are some remarkable facts known in reference to these bodies. First, *they are all round or globular bodies*. The Sun, Mercury, Venus, the Earth, Mars, and all the rest, are globes. Why should this be so? Why not triangles or cubes or pentagons? Or, if round, why not round like a cylinder, a table, a top, or a sugar-loaf? Or why not one round, another square, and a third flat? We see a resemblance among them, which becomes more apparent when we are better acquainted with them.

Not only are they all globes, but *they are all in motion, and moving in one direction, and nearly in one plane.* The sun does not move around the earth, as the ancients supposed that it did; but it does revolve upon its axis from west to east. In the same direction, Mercury travels around the sun; so Venus, so the Earth, and all the primary planets. These bodies, we see, are no longer lone wanderers in space, that chance has scattered abroad: they are members of one grand family, bearing so great a resemblance to each other, that we at once suspect that they must have had a common origin.

Another remarkable fact connected with the planets revolving around the sun, and the satellites, or moons, that revolve around them, is, that *their periods of revolution are increased in proportion to their distance from the bodies around which they revolve.* Thus Mercury travels around the sun in 88 days, Venus in 224 days, the Earth in 365 days, Mars in 686 days, Ceres in 4 years and 9 months, Jupiter in 12 years, Saturn in 29½ years, Uranus in 82 years, and Neptune in about 164 years. So that, if Neptune is blessed with seasons, it has 41 years of spring, 41 of summer, 41 of fall, and 41 of winter. I pity the child born on that planet about the end of the fall, with a forty-one-years' winter ahead of him, especially such a winter as Neptune's must be. Should he live to be one year old, he would be considered a marvel of old age in this faster-going planet of ours.

Saturn's first or nearest satellite performs its revolution around it in 22½ hours, the second in 1½ days, the third in 2 days, the fourth in 2½ days, the fifth in 4½ days, the sixth in 16 days, the seventh in 22 days, and the eighth in 79 days. Jupiter's first satellite revolves around it in 42 hours, the second in 3½ days, the third in 7 days, and the fourth in 16½ days.

One more important fact has been observed, — *the sun and all the primary planets that have satellites accompanying them rotate upon their axes in less time than the planets or satellites revolve around them.* Thus the sun revolves upon its axis in 25½ days, — a much shorter time than that of the revolution of any planet around it; the earth revolves upon its axis in one day; while the moon requires more than 27 days to describe a revolution around the earth. Jupiter revolves upon its axis in 10 hours, while its nearest satellite revolves around it in 42 hours. Saturn's time of rotation is 10½ hours, while its nearest satellite requires 22½ hours to revolve around it. These facts, and others of less importance, have led philosophers to inquire for a cause adequate to account for them; and thus the Nebular Theory had its origin.

According to this theory, there was a time in the almost infinite past, when all the matter now contained in the sun, the various bodies of the solar system, and the stars, existed as nebulous or gaseous matter, intensely heated, and exceedingly rare. In this, as it cooled, centres of aggregation were formed, toward which flowed the nebulous matter; producing by that flow rotary motion in the aggregated centres or immense suns thus formed. In this way, the grand centre of our solar system was produced, occupying all the space now occupied by the planets, and much more, — a globe, probably, a hundred thousand million of miles in diameter, and in an exceedingly heated condition. This stupendous globe was in motion from west to east, and radiating its heat into space, which is known to be more than two hundred degrees below zero. As its heat decreased, it necessarily became smaller, cooling bodies, as a rule, contracting in size; and, becoming smaller, moved with greater velocity, on the same principle that a stone fastened to a string, and whirled around the finger, flies with more velocity as the string winds around the finger and becomes shorter. But, as the velocity increased, the centrifugal force heaped up the matter around the equator of the revolving sun, and caused it eventually to separate from the main body in the shape of a zone, or ring. By experiment it has been proved that a drop of oil floating in alcohol will do this when revolving with sufficient rapidity. The matter composing this ring, on breaking up, which, unless exactly balanced, it would be sure to do, flowed into a globe by virtue of the attraction of gravitation; the same law that rounds a dewdrop on a grassy spear sphering the worlds in space: for these grand laws know no great and no small, — just as ready to round a world as to round a rain-drop.

Thus we may suppose Neptune was launched into space; at first a nebulous ring, and finally a rotating globe or planet, the first-born of the Sun; it, in turn, becoming the parent of smaller globes or moons. Thus, also, Uranus and her moons; Saturn with a family of eight satellites and rings, so nicely balanced, that they have been preserved, which, by their very exceptional appearance, favor this theory of their origin; and, lastly, the Earth, with her solitary child the Moon, the last fertile planet of the solar system.

How beautifully this theory accounts for the facts of

the solar system that have claimed our attention! Why are the planets round or globular bodies? Because they were originally fluid; and the law of gravitation rounded them in the same way that it spheres a tear-drop as it falls from the eye. Why are the planets moving bodies? why do they all move in the same direction, and nearly in the same plane? and why is that plane nearly coincident with the sun's equator? The sun moving from west to east originally, as it now does, communicated its motion to the successive rings that were separated from it; and, as they were separated from the sun's equator, they still remain in about the same plane. We see, too, why the outermost planets and satellites take the longest time to perform their revolutions. When Neptune was separated from the sun, the sun was of immense size, and moved slowly; but as it became smaller, and moved more rapidly, every planet partook in regular order of this increased rapidity of motion: and thus Mercury travels round more rapidly than any other planet; and, for the same reason, the nearest satellite of any planet is the most rapid, and the sun and planets more rapid in their revolutions than any body revolving around them.

The present condition of some of the nebulæ seen in the heavens, which cannot be resolved into stars by the best telescopes, indicates the existence of systems in that nebulous condition from which this theory supposes our solar system originally came. Nay, recent spectrum analysis has demonstrated the gaseous condition of some of the nebulæ; while it shows that the sun is in such a heated condition, that iron, copper, zinc, and other metals, exist in its atmosphere in the condition of vapor; while the telescopic appearance of the moon, with its mighty craters, its lava-streams, and marks of igneous action nearly everywhere, indicates an original fiery condition for it, such as we have proved to have been the state of the earth at an early time, and such as is demanded by this theory.

The Nebular Theory was first suggested by the elder Herschel, was elaborately developed by La Place the great French astronomer, and has received the support of the best scientific minds that have investigated it. Mitchell, the Cincinnati astronomer and lamented patriot, says in reference to it, " Such is a brief outline of one of the most sublime speculations that ever resulted from the efforts of human thought." Dr. Mantell, an English geologist, says, " This theory of the condensation of nebulous matter into suns and worlds, marvellous as it may appear, will be found on due reflection to suggest the only rational explanation of the phenomena observable in the sidereal heavens and in our own globe, according to the present state of the physical sciences." Dr. Buckland, in his Bridgewater treatise, remarks, " The Nebular Hypothesis offers the most simple, and therefore the most probable, theory, respecting the first condition of the material elements that compose our solar system." Dr. Pye Smith says, " The Nebular Hypothesis, ridiculed as it has been by persons whose ignorance cannot excuse their presumption, is regarded by some of the finest and most Christian minds as in the highest degree probable." Lastly, Humboldt, the noblest name of which science can boast, says in his " Cosmos," " In the first formation of the planets, it is probable that nebulous rings revolving around the sun were agglomerated into spheroids, and consolidated by a gradual condensation proceeding from the exterior towards the centre."

Behold the earth, then! —

> "It goes its glorious course to run,
> A fire-globe whirled from the burning sun."

Uproll the curtain that unnumbered ages have dropped, and view the wondrous scene. Before us spreads a fiery ocean, bounded only by a fiery sky: its lightning-capped billows heave heavily under the influence of sun and moon; and now, as if mad, they leap in fury to the ruddy clouds that lower above them. Hissing, seething, boiling like a huge caldron, while dense vapors rise continually from its agitated surface, it presents to us a picture that none but a demon could truly paint. Its air, if air it may be called, is hotter than the volcano's breath, and more deadly than the dread simoom. There is no night, with grateful shade and cooling dews; no winter, whose piercing winds may assuage this terrific heat: there is but one unvarying, fiery day; one interminable, burning year. Surrounded by an atmosphere whose bulk is vastly greater than that of the earth itself, it looks more like a blazing comet than a world destined to be the peaceful abode of human beings.

In this state of the earth, all the metals, in consequence of the intense heat, must have existed in the atmosphere in the condition of vapor; and, as it cooled, they dropped upon its surface in showers, — gold, silver, copper, iron, lead, tin, and quicksilver, — lakes of these, rivers of these, rising in vapor, forming clouds, and condensing over and over again ages before any water existed on the globe. Where was the water during this period? All the water now contained in rivers, lakes seas, and a component part of rocks, was then in the atmosphere. "But the atmosphere is not large enough to hold it," replies an objector. True, it is not now, but amply sufficient then. The oxygen of the globe (half of the earth's known crust being composed of it) was then in the atmosphere; all the carbon now locked up in the coal-beds (coal containing from fifty to ninety per cent of it) must have been there also as gas; the carbon in limestone was also there (it composes one-eighth of all limestone rocks); united with oxygen, in the form of carbonic-acid gas, every cubic yard contains seventeen thousand cubic feet. What a volume in the thousands of cubic miles of limestone! and what an immense atmosphere there must have been before the consolidation of so large a part of it into rock! Thus the vapor of all the water in the world could be easily contained in it.

It is exceedingly difficult to obtain a correct idea of the condition of the earth during this time, — age after age, heat radiating into cold space; rock forming on the surface of the shoreless, fiery sea, as forms the ice on a lake; forming and breaking and re-forming, only to rend again, as the red billows sweep beneath, to unite in firmer bands than before. Sluggish waves incessantly roll round and round the globe; and these, breaking, and heaping up the black sheets of rocks as they form upon the surface, produce craggy islands, against whose banks the waves dash, and congeal into rock. Islands thus formed unite into continents, and at length cover the earth's face, and hide its glory: it ceases to shine in the heavens a star, and must henceforth be indebted to other bodies for its light, and eventually for heat.

Let the earth be gradually heated until it assumes the condition of vapor. Not long, and the tropical regions would be uninhabitable by man, and there would be a general migration to the north. The ice would melt from Greenland and the north and south polar regions, and ere long they would abound with living beings now only found in temperate climes. The whale, seal, walrus, and white bear, would make their way to the poles; and the Esquimaux would follow them. The heat would become so great eventually, that, in the temperate regions, the sheep, the ox, and the deer could no longer exist: one by one they would perish as an extra tropical climate spread over the whole earth. The heat increasing, man dies, and the beasts generally perish; those living in trees and in extreme northern regions remaining longest. Then the birds die: but insects and reptiles swarm in increased numbers the world over; but, the heat continuing to increase, they at length also die. The plants of the temperate regions are all gone, and even the tropical plants at last; and a desert world is here. The water still abounds with fish, and sea-weeds flourish in a myriad forms; but these at length expire. Infusoria exist till the water is boiling hot; but at length they cease to exist, and all life is gone. In time, all the water is turned into steam. The heat continues, and the oil in the earth is made to boil: in some places it takes fire and burns, and in others is driven off in vapor. The coal burns at length, and its carbon is transformed into carbonic-acid gas. The limestone of the world is burned into lime; the metals melt; quicksilver rises in vapor, then lead, copper, iron, and gold; the mountains sink and dissolve; and, as the heat increases, the rocks themselves are resolved into gases and vapor, and ascend: the world has become a mass of molten matter, belted by an enormous atmosphere; and we have at last an unseparated nebulous mass, like that from which all we behold has been produced.

Some faint conception of the time consumed in forming a foundation for the continents to be built upon may be obtained by considering the following facts: Lava from Mount Ætna, in 1819, moved down a considerable slope, at the rate of a yard an hour, nine months after its emission; and in the crevices of the mass a dull-red heat could be seen by night, and vapor rising from them was visible by day. Lava from the same mountain at a previous eruption was in slow motion ten years afterward. A mass of lava sixteen hundred feet thick was ejected from Jorullo, in Mexico; and, though a hundred years have passed, it is not cool yet: travellers push their walking-sticks into the crevices; and the heat is still intense enough to char the ends of them. How long must it have taken the earth to cool sufficiently to allow a rocky crust to rest upon its surface! Count ten thousand years for every hair of your head, and you have by no means over-estimated the time necessary. Yet it came; for Nature commands eternity, and she is never niggardly of time. Now is the floor of the world laid down, black, hard, bare as the pavement of the street, hot as an oven: it stretches away, a craggy desert of desolation surrounding the globe; its monotony broken only by the numerous volcanic mouths that pour out lava-floods continually.

The time comes when this is cool enough to allow the water to condense upon it; and, as it falls in drops, it is speedily dissipated in steam, rising in dense clouds to the heavy atmosphere. This cooled the earth's surface more rapidly than before; and at length the water remained in the hollows. The oceans were born, boiling, dark, heavy, and impure, amid heavings, roarings, flashings, and terrible storms and convulsions. There stretches a chasm miles away; and between the dark, craggy walls that bound it, see the boiling lava, like a river of melted gold, rolling its waves along! Now water flows into it: the ground moves beneath our feet. What an explosion! All the gunpowder of the world fired at one blast could not equal it. Rocks in mountain masses are flung into the air like pebbles, and, falling, break through the yielding crust, and disappear in the boiling abyss beneath. The walls of the chasm are now uniting: their pressure forces up the lava in fiery fountains. Fire and water are struggling for the mastery with the wide world for an arena. Which shall conquer? Turn and view that muddy ocean behind you; its turbid waves dashing against the rocky shore: dense, steamy clouds hang over it by day and night, whose darkness is only relieved by the numerous fires that cast a lurid glare on the vapory envelope of the globe.

GRANITIC AND METAMORPHIC PERIODS.

AGE OF MINERALS.

It was during this stormy period that the great foundations of the globe were laid, including the time when the granitic and metamorphic rocks were deposited, and before life had made its appearance, — a period sometimes called the azoic age, or the age destitute of life: *a*, in Greek, from which this word is derived, meaning "without;" and *zoe*, "life."

Were I asked of what books are composed, I might say, Of chapters, or of sentences or words or letters (of which there are twenty-six); these making by their combinations the hundred and twenty thousand words of our language. So, if asked of what the earth is composed, I might say, Of various formations, or of rocks and metals, or of minerals (of which we have about seven hundred), or of elements (about sixty of which have been discovered). These, by their combination, make minerals, as letters make words; minerals make rocks, as words make sentences; and rocks compose the various geologic formations. Of these sixty elements, some are quite rare, and many are only known to the chemist; but about twelve of them are of practical value, and the student of geology needs to be acquainted with them.

Half of the earth's crust is *oxygen*, — an invisible gas. It was discovered by Priestley in 1774. It is the supporter of life and combustion. Water contains nearly ninety per cent of it; the atmosphere, more than twenty per cent. Sand is more than half oxygen; and limestone and clay, about half; one-half of the globe, as we are acquainted with it, made of a gas that no man ever saw, tasted, or smelt!

One-quarter of the earth's crust is *silicon*, — the base of silica, quartz, sand, and flint; which are silicon combined with oxygen: when thus combined, it is the principal ingredient of all rocks except the limestones. In pure water, silica cannot be dissolved; but when it is very hot, or when it is warm, and contains potash or soda, it is readily dissolved, and may then be deposited wherever the water flows. The pores of organic bodies buried in the earth have in this way frequently been filled with silica, and the shape and general appearance of them preserved for ages.

Another important element is *aluminum*, — a white metal between tin and iron in many of its qualities, but light as chalk. United with oxygen, it forms alumina, which is pure clay. Common clay, of which brick is made, is alumina containing impurities. When pure alumina is crystallized, it forms the sapphire, next in hardness to the diamond.

Calcium is a light-yellow metal, a little harder than lead, and readily unites with oxygen, forming quick-lime. Limestone is the carbonate of lime, formed by the union of carbonic acid and lime. When put into the kiln, the carbonic acid is driven off by heat, and quick-lime is left. One ton of good limestone yields about eleven hundred weight of lime. When lime and sulphuric acid unite, the result is sulphate of lime, or gypsum, found in immense quantities in some parts of the globe. With phosphoric acid, lime forms phosphate of lime, the principal ingredient in bones.

Magnesium enters quite largely into the composition of some of the rocks. It is a white, brilliant metal, readily uniting with oxygen; when it forms magnesia. This, united with silica, forms hornblende, talc, soap-stone (which we now make into stoves), and serpentine.

Iron forms nearly two per cent of the earth's crust. It is very seldom found in its native condition, except in meteorites; but is found combined with oxygen, carbon, and sulphur. The readiness with which it combines with oxygen may be seen in the rust so easily formed on its surface; rust being oxide of iron. It is very abundant, and widely diffused; being found in nearly all rocks and minerals.

Potassium is a silver-white metal, so soft that it may be moulded like wax. Its affinity for oxygen is so strong, that, when thrown upon the surface of water, it decomposes: the water unites with the oxygen, forming potash, and liberates the hydrogen, which burns with a beautiful flame. Potassium exists chiefly in felspar and clay, and constitutes about five per cent of the igneous rocks.

Carbon is generally found united with oxygen in the form of carbonic acid. In this condition, a vast amount is locked up in coal and lime. The atmosphere contains about one part in twenty-five hundred by weight of this gas; but, before the coal and limestone beds were formed, it must have contained vastly more. Charcoal is nearly pure carbon; and the diamond, so utterly unlike it, is carbon absolutely pure and crystallized. It is estimated that nearly two per cent of the earth's crust is carbon.

Hydrogen, the lightest known substance, is another somewhat abundant element. It forms one-ninth part of water; and since water enters largely into the composition of rocks, gypsum containing twenty per cent and serpentine thirteen per cent, hydrogen, in this way, constitutes about half per cent of the earth's crust.

Sodium, a light metal, and *chlorine*, a heavy gas, constitute salt; and are therefore widely distributed upon the surface of the globe, and beneath it. Sodium, in the form of its oxide (soda), is contained in the igneous rocks generally. Granite contains about seven per cent.

Lastly, *sulphur*, which is sometimes found in a perfectly free state, but more frequently in combination with the metals, as iron, lead, copper, &c.

Of these elements, nearly the whole earth, as far as we are acquainted with it, is composed. By union, they form the various minerals: thus silicon and oxygen make *quartz*, which, ground to powder, gives us sand; and this, united again by pressure, makes sandstone. Flint is a variety of quartz. In its various forms, quartz constitutes one-half of the earth's crust.

Silica, uniting with alumina (the oxide of aluminum) and potash (oxide of potassium) and soda (oxide of sodium), forms *felspar*, which constitutes one-tenth of the crust of the earth. *Mica* is composed of nearly the same materials in different proportion.

Limestone, or the carbonate of lime, which is formed by the union of carbonic acid (carbon and oxygen) with lime (calcium and oxygen), forms about one-seventh of the earth's crust.

Hornblende, a black or greenish-black mineral, which enters into the composition of most of the igneous rocks, is composed of silica and magnesia (magnesium and oxygen). Hornblendic rocks are very tough, and dark in color. *Talc*, a soft mineral, sometimes called soapstone, having a greasy feel, and *chlorite*, a mineral somewhat like it, are composed also of silica and magnesia.

Most of these minerals must have existed on and in the earth at an early period, and long before life made its appearance.

Fig. 1* is a representation of the various rocks composing the earth's crust, in the order in which they were deposited. The lowest is granite, generally supposed by geologists to have been produced by the cooling of the original fluid mass of the earth, and hence the great underlying rock of the globe,— that on which all the others have since been deposited, supporting all the varied geological formations, as the rock or soil supports the houses that rise above it. Dig where we please, there is good reason to believe, if we could only go deep enough, that we should eventually come to the granite. Strange to say, though the granite is the lowest rock of all, it is frequently found on the tops of the highest mountains. Thus granite is found on the summit of Mount Washington, the highest mountain in New England; on Mont Blanc,

* See Frontispiece.

the loftiest mountain in Europe; and many peaks of the Rocky Mountains. The granite found on the tops of these mountains doubtless lay at a great depth beneath the surface at one time, but has been heaved up, through the rocks that lay above it, to the position that it now occupies; while the overlying rocks were broken through and laid upon the sides of the mountains thus formed: and hence we frequently observe rocks on the surface that were once miles beneath it, and obtain a knowledge of the earth's interior which is most valuable.

Granite is a rock well known in New England (where it is quarried in numerous places), and generally in the north and north-east of the United States, by numerous bowlders, which lie scattered over the country. The color varies much in different specimens,—in some, gray; in others, white or flesh-colored. It has consolidated gradually under immense pressure, and is invariably crystalline in its structure: the crystals sometimes are found several inches in diameter, but more frequently not more than an eighth or a quarter of an inch. True granite is composed of felspar, quartz, and mica. Quartz is the hardest portion of the granite, and will readily scratch glass. It has a bright, glassy appearance, and breaks into angular pieces under the hammer.

Felspar is not quite as hard as quartz, nor yet as bright, and is more readily decomposed. It is generally found of a yellowish-white color, and is the commonest of all substances except quartz and iron. When decomposed, it forms a white clay, which is used in the manufacture of porcelain and china-ware. It is known in China by the name of *kaolin.* Upwards of twelve thousand tons are carried every year from Cornwall to the English potteries.

Mica, the third ingredient of granite, improperly called isinglass, is bright and shining like silver; and is so soft, that it may be scratched with the finger-nail. It can be readily separated into thin plates; and in Siberia, where it is obtained of a large size, it is used for windows, instead of glass. Sheets more than two feet in diameter have been found in the granite of Acworth, N.H. It is frequently used in the doors of stoves; for the heat does not crack it, and we can see the cheerful firelight through it. It has been used for windows on board of men-of-war, as it will not break by the concussion of the guns.

Felspar generally forms the principal part of the granite. Sometimes mica is absent, and hornblende takes its place: the rock is then termed Sienite, from Syene in Egypt, where it was supposed to abound. When mica is also present, it is termed Sienitic granite. When felspar is the principal ingredient, the quartz and mica being very rare, it is called felspathic granite. Many minerals are found in granite; but it will not be necessary for me to refer to them.

As the water descended, it wore down this granitic crust ridged up by the cooling of the earth, just as rivers wear down rocks to-day,— more rapidly then, however; for the water was not an honest combination of oxygen and hydrogen as it is now, but more like sulphuric acid, and its dissolving power was much greater than water. The sediment carried down was deposited in the hollows of the ocean, and here a new class of rocks was formed.

As two grand agents have been concerned in bringing the world to its present position,— fire and water,— we have, as the result of their operation, two classes of rocks,

—the fire-made and the water-made; but, beside these, we have a third class, formed by water, and modified by fire. These are called *metamorphic* rocks, so called from the Greek word *metamorpho-o*, which means to transform; these rocks being changed from their original form. At first, laid by water at the bottom of the ocean as sediment, they have a stratified appearance, such as belongs to the water-made rocks: but, the heat passing into them from the underlying fire-made rocks, in many cases they have become melted, and in others very much heated; so that, on cooling, they have crystallized, and hence have the appearance that belongs to the crystalline or fire-made rocks.

Where the granite was worn down, and carried into and laid at the bottom of an ocean whose waters were disturbed, so as to mix the material up, the sediment, forming in time into rock, would contain all the ingredients of granite, but in a finer form, and deposited in layers. Such a rock is gneiss; very abundant in New England, among the Rocky Mountains and most mountainous regions.

The quartz of the granite ground to powder would be sand; this, separated from the other ingredients of the granite, and deposited in an ocean whose waters were undisturbed, would by pressure make sandstone; and this sandstone, heated and cooled, forms another metamorphic rock, called quartzite.

The felspar, operated upon in like manner, would be formed, first into beds of shale, and then metamorphosed into slate. The slate with which we cover our houses, and on which our children write at school, was, it is believed, produced in this way from the felspar of the granite. The mica, in a similar manner, formed the mica schists or mica-slates which make such beautiful flagging for our streets.

During the time that the rocks of these two formations were deposited,—the granitic and metamorphic,—it appears that no life existed on the globe; no bird stirred the heavy air with its wing; no beast trod the heated ground; no fish occupied the steamy waters: but the grand old world was marching steadily on, and the ground was being prepared, from which fruitful harvests in future ages were to be gathered.

There was a time when the granitic and metamorphic rocks covered the whole globe: on them the waters rested, and of them the land-surface was everywhere composed. At the present time, however, but little of this primitive surface remains. It has been covered by sediment, carried down or worn down by the waters; and only those portions exist which have not been worn away or under water, and so have never been covered by sediment. Or where we find these rocks on the surface, in mountain-regions, they have been elevated from beneath the overlying rocks; while, in still other places where they are found, the rocks that once lay above them have been swept off by denuding agencies operating for unnumbered ages.

In Scotland, they cover a considerable portion of the Highlands and islands. The north of Ireland, and the Wicklow Mountains in the east, the Cumberland and Cornwall Hills in England, the Alps in Switzerland, and the Pyrenees in Spain, are composed of them. Large portions of the mountain-regions of Asia, Africa, and America, are composed of rocks belonging to this age of minerals.

The most elevated portions of the Rocky Mountains

are composed of metamorphic and granitic rocks. The backbone of the future continent — the present line of that mountain-range — was probably indicated at a very early period.

In Canada, we find a belt of this rock, extending from the St. Lawrence, near the outlet of Lake Ontario, in a north-eastern direction to Labrador, and in a north-western direction to the Arctic Ocean; and this seems to have formed the nucleus of the eastern part of the North-American continent. Slowly elevated, as the waters of the ocean were slowly drained off by the deepening of its hollows, there were deposited around this, as a centre, the younger formations, in the order of their age; so that we can behold the layers of the continent's growth, as in a tree we see the rings which mark its yearly development.

LECTURE II.

If a man should commence to study astronomy with the idea firmly fixed in his mind that the universe is but a few thousand miles in diameter, how futile would be his endeavors to master the grand truths of this magnificent science! With such a contracted notion as this, he could form no adequate idea even of the standpoint from which he observed the infinite heavens. So in geology: let a man commence to study it with a firm persuasion that the earth is but a few thousand years old, and what can he ever learn of geology? He has not room even for the titlepage of the mighty volume spread before him: he cannot even account, in such a scanty period, for the soil that lies under his foot. An enlarged conception of the element of time is absolutely essential to a solution of the geologic problem.

We call this a grand old world, with but little idea, sometimes, of the great significance, in this connection, of the words we employ. We call men old when age has whitened their locks, unstrung their sinews, and graven their faces with those unmistakable lines which no art can erase. We call the oak-tree old whose giant trunk, time-scarred and lichen-browned, stands in majesty upholding the stalwart branches from which its myriad twigs their leafy banners wave. We speak of old English castles, whose ivy-crowned ruins have inspired generations of poets, whose towers looked down on the feudal barons as they marshalled their vassals for the deadly fray, and whose halls rang with the wassail revels of the olden time. The Pyramids of Egypt we say are old, — those mighty monuments of a past civilization, that stood on their firm bases when Britain was an island inhabited only by savages whose dress was the paint that besmeared their bodies, or the skin torn from some wild beast. Yet what is the age of these compared with the age of the world? They are the veriest babes of time, the ephemera of a summer's day: they resemble the bubbles that float on Niagara's stream, glittering for an instant on its turbulent breast, then disappearing forever.

Let us look first at the soil, the earth's covering, and the youngest thing with which geology deals. A tree was cut down in Calaveras County, California, that measured thirty-two feet in diameter. There were thirteen rings of annual growth to an inch of it; so that the tree was 2,496 years old, — a tree that was a sapling when Nebuchadnezzar was a boy, that was nearly two hundred years old when Socrates was born. And yet this tree is by no means the oldest vegetable monument on the globe. A yew at Fortingall, in Scotland, is calculated to be 2,600 years old; and one at Braburn, in Kent, 3,000. There is a baobab-tree at Senegal, in Africa, in which an incision was made, and the concentric rings counted; and from that it was calculated to be 5,150 years old. Yet there is a cypress in Chapultepec, 117 feet in circumference, which Humboldt considers still older.

The soil must have been there before these trees began to grow, or how could they have taken root? The parents of these trees must have had soil likewise for their growth. Who shall say how many generations of such trees are represented by this soil? what ages were spent originally in pulverizing the hard rock to make it? What historian shall write the biography of these twelve inches of earth? It is older than Britain and the Druids, older than Gaul and the Celts. Greece the ancient, and her gods, are young compared with it. Older even than India, the grandfather of nations, away into the shadowy past stretches the soil we so carelessly tread under our feet; yet the age of the soil is but one of the days of the earth's vast generation.

Beneath the soil, in all northern countries, lie beds of sand, gravel, or clay; and mixed with these, or lying above them, are rocks known by the name of bowlders, many of which have travelled hundreds of miles from the places they originally occupied. Underlying the soil, they were, of course, laid down before it. Some of the beds of gravel and clay belonging to the drift, as the formation is termed in which they are found, are, in places, hundreds of feet in thickness, indicating the immense period of time occupied in their formation. In some of these beds, in Great Britain and Europe, have been found the bones of enormous oxen, nearly as large as elephants; bears larger than the great grisly of California; monstrous hyenas and elephants, belonging to species now no more. During a portion of the time that these beds were being deposited, glaciers thousands of feet in thickness, and hundreds of miles in extent, moved over the face of North America, Northern Europe, and Northern Asia, grinding down the rocks, and transporting vast quantities of material to distant localities. During this time, also, the sea occupied much of what is now dry land; and gigantic icebergs, laden with rocks and *débris*, went sailing toward the south, and, upon melting, deposited their burdens on what was then the floor of the ocean, — now much of it land occupied by man.

Below these drift or glacial beds lies the tertiary formation, consequently older still. It consists of beds of clay, sand, gravel, marl, sandstones, and limestones; having a thickness, in some places, of thousands of feet. Some of these beds were formed at the bottoms of lakes; others in estuaries, where salt and fresh water intermingled; and some at the bottom of the ocean. In them have been found, in Great Britain, multitudes of strange fruits, and remains of plants; indicating the existence of a tropical climate there during the period of their growth, which is further evident from the character of the shells associated with them, many of which belong to genera whose living representatives are found only in tropical seas. In the tertiary beds of Germany are large deposits of brown coal, containing trees, in one of which seven hundred and ninety-two rings of annual growth were counted. Gigantic turtles and tortoises, as well as crocodiles, have been discovered in some localities, and the remains of many extinct beasts: the gypsum quarries of Paris, in rock belonging to this formation, abound with their bones.

We have now travelled back so far that we are in a new world, stranger to us than America was to Columbus. The Alps, Himalaya, and Andes lift their heads but a few thousand feet above the surging waves; while Ætna lies slumbering beneath the bed of the

Mediterranean, then covering a large portion of Northern Africa, and occupying twice the space it does at the present time. Since the mountain-chains determine the position and size of the rivers, how different is the whole face of the land! The ocean covers the spot now occupied by the largest cities; and immense sea-monsters disport themselves over millions of square miles now occupied by man. One-third of England was covered by the sea, half of Ireland, and three-fourths of Russia, a large part of North and South Carolina, the whole of Florida, a large portion of Louisiana and Mississippi, all the eastern portion of Texas, and a great part of Alabama and Georgia. It might be supposed that we had by this time arrived nearly at the end of our journey: but we have, however, only fairly commenced our travels into the mighty past; and untrodden realms lie still before us.

Beneath some of the tertiary beds lie beds of chalk: hence we discover that the chalk or cretaceous formation is older than the tertiary. From the materials of which it is composed, we learn that it was deposited at the bottom of the ocean; and from its thickness, being with its accompanying beds of clay and greensand nearly two thousand feet thick, some idea may be formed of the vast period during which it was in process of deposition.

Of all the existing animals, not one has been able to accompany us in this tremendous journey: all we find are new; the familiar islands, rivers, and continents are gone, or so strangely altered that we no longer recognize them; and we are indeed strangers in a strange land.

Cropping out from beneath the chalk in England,

and covering a considerable portion of the eastern part of the island, are a succession of beds of limestone, marl, shale, and sandstone, sometimes termed the oölitic formation. These beds, in the aggregate, are more than two thousand feet in thickness; and since, in places, they underlie the chalk, were evidently deposited before it.

In them we find the remains of gigantic lizards, that then crawled over the face of the world; one generation after another flourishing, dying, and becoming entombed in muddy deposits, which eventually hardened into rock.

Below them are found beds of red sandstone, magnesian limestone, salt, gypsum, and marl; some of the beds of salt alone being hundreds of feet in thickness. In some of the sandstones are found impressions of the feet of what are supposed to be birds, and reptiles; and some that seem to unite the characteristics of both, evidently differing considerably from all existing animals. Farther into the domain of ancient Time we are marching; but we are far from the beginning yet.

Below these beds are the coal-measures, and underlying limestones and sandstones, representing by their enormous thickness, in some places from ten to fourteen thousand feet, the great space of time occupied in their deposition. So far have we advanced, that all beasts, and even birds, have perished by the way: one by one they have drooped and dropped and died; and nothing is left on the land to accompany us on our journey but insects and a few frog-like reptiles that are still to be found hopping over the face of the steaming ground.

Below these come the Devonian beds of sandstone,

limestone, and conglomerate; in Scotland, ten thousand feet in thickness, and, in some parts of the United States, not less; its sandstones, in Scotland, crowded with mailed fishes, and its limestones composed of the remains of shells, corals, and other marine animals. What vast ages must be represented by their accumulated remains!

Beneath the Devonian formation we find the Silurian, attaining, in England and Wales, a thickness of from twenty to thirty thousand feet, composed of rocks laid down at the bottoms of ancient oceans, and many of them as full of the remains of old forms of life "as a straw-stack is full of straws." What time spent in building these mountain-monuments of the dead!

In the neighborhood of Perth, Can., are to be seen the rocks of the Laurentian formation, several thousand feet in thickness, in which a few obscure fossils have been found. With the lowest of these beds, all life has vanished; and we march into the night of the still farther past, lighted but by the fitful glare of dread volcanoes, all alone.

Then follow the metamorphic rocks, whose thickness, on the flanks of the Andes, has been estimated to be from ten to fifteen miles; the sediments of oceans boiling hot, worn from the pre-existing granite by the agency of water, — that granite formed during ages numberless from the cooling of the original igneous mass of the globe. What an immense period! — "an eternity, all but the name."

Having obtained some faint idea of the vast space of time during which our planet was in process of formation, as well as the method of its advance, — without which we should grope our way most blindly without a guiding star, — we can now walk confidently along its pathway, feeling the ground at every step, as we consider, in ascending order, the construction of the various geological formations, and see how, out of the original chaos, came the order and beauty that we behold to-day.

The metamorphic period was that on which my last lecture closed, — a period indicated by those immense beds of gneiss, mica-schist, talcose-schist, slate, and conglomerate, which manifest the long ages of desolation during which the lifeless globe swung round the sun, apparently in vain. Our planet was then a world of belching volcanoes and spouting geysers, of heaving earthquakes and howling tornadoes, — a land on which the sun had never shone, unblessed by the smile of the moon, unvisited by the light of a star. Sulphurous clouds canopied it, from which fell dark showers; while black torrents poured over craggy rocks into turbid oceans that lay boiling beneath them.

The rocks formed during this period are destitute of all signs of life: but those lying immediately above them contain what are called fossils; and these are of the greatest importance to the geologist. What are fossils? Animals or vegetables buried in the earth by natural causes, and preserved; or any indications of their existence found in the rocks. They may be shells almost unchanged, as these post-pliocene shells from South Carolina; or bones, like these from Colorado, somewhat heavier than they were originally, by the infiltration of earthy matter into their pores; or wood changed into flint-like stone, as in these beautiful specimens from the miocene beds of Texas, with which I can strike fire just as readily as with two flints. In them the structure of the wood is just as plainly visible as in wood just detached from the living tree. The silica, in similar specimens, has been removed by steeping them in hydro-fluoric acid, and the woody fibre found so well preserved, that it could be used in determining the genus of the original plant. I have found specimens partly converted into stone; the rest lignite, still retaining much of its original woody character. Wood has been found in a Roman aqueduct partially converted into stony matter; and the stave of a cask which had been in the well of the Castle of Gotha for a hundred and fifty years was so petrified by the oxide of iron as to take a polish by friction.

A fossil may also be the cast of an animal or plant. Sometimes, when shells are buried at the bottom of the ocean or lakes, mud is pressed in, or filters in, and fills the cavity of the shell. This mud, in time, hardens into rock; while the shell on the outside decays, and is removed by infiltration. The cast of the inside of the shell thus formed is a fossil as truly as the shell itself; and in some rocks they are the only fossils we find. In mud-banks on the coast of Nova Scotia, I have seen multitudes of such fossils in process of formation, — the mud in the shell half hardened into rock, and the shell partly decayed. Similar specimens, less advanced, I have also seen dug out of the Erie Canal, New York; also brought up by a dredging-machine at St. Joseph, Mich.

In some cases, animals have walked over mud; and this, hardening into rock, has preserved the track made upon it, producing a fossil footprint. The marks of rain-drops, ripple-marks, and sun-cracks, have been preserved in a similar manner; and, though not organic, they are of the greatest service in reading aright the past history of the globe.

These fossils are the letters in which the history of the world is written; and without their assistance we should not have known the past history of our planet probably for ages to come.

By them we decide the comparative ages of rocks, — a matter of the greatest importance. Certain forms of life lived at certain times, and their remains became buried in the sediments deposited during those times: hence, when we find rocks containing fossil-forms with which we are acquainted, we can refer those rocks at once to that period in the world's history when we know those forms of life existed. Thus in the rocks at Cincinnati, O., at Ottawa, Can., at Trenton Falls, N.Y., and at Dudley, Eng., we find similar fossil-forms, and refer the rocks to the Silurian formation; for the animals whose remains we find in them lived at that period, and no other. Rocks that are separated by many thousands of miles are thus known to be identical in age, or nearly so; and, the farther we go back in time, the more certain we may be of this, the general condition of the earth and ocean being similar over the entire globe at an early period in its history.

When we find in certain beds the remains of land-plants, of fresh-water shells and fishes, and of land-animals, we may feel sure the deposits were made in a lake of fresh water; but if we find strata containing land-plants and sea-plants, marine shells and bones of land animals, mixed confusedly together, we may know that we are examining a deposit made by a river pouring sediment into the sea. When all the forms are marine, which can readily be known, they, in like manner, speak of the condition of their formation; and we can even tell whether the sea was deep or shallow by the prevalence of certain forms.

Nearly all the rocks above the metamorphic contain more or less fossils; and in many parts of the country they are almost exclusively composed of them: this is especially the case with limestones. About twenty thousand fossil-shells, which are the most readily preserved of all bodies, are already known; and that number will doubtless be more than quadrupled when the crust of the earth has been thoroughly examined.

By fossils we learn that the earth has been inhabited by living beings for millions of years, rising higher and higher in the scale; that death has swept away individuals and species, but only that life might supply new individuals and more advanced species. They enable us to answer a thousand questions that the inquisitive soul of man is ever asking; and, when they do not fully answer them, they yet give much light and satisfaction. Where did life begin? On the mountain-top, among the silent clouds, and, descending, spread over the newly-formed valleys? by the river-side, in the slime left by its overflowing waters, so well calculated to nourish the vital germs? on the undulating surface of the land, warm and steaming from the internal fires and descending torrents? Or did it commence in the world-embracing oceans, whose heaving waters laved the dark islands that dotted its wide surface? And *what were the first organized existences?* Pines wagging their heads on the hill-tops? fruit-trees bending beneath their blushing load? grass carpeting the valleys with velvet, beautifying the earth previous to its occupancy by its grand controller, man? Were elephants the first occupants, shaking the ground with their tread? or timid rabbits peeping from their burrow? whales in the ocean, or minnows sporting in the brook? Had it been left to man's imagination to decide, we should probably have had monsters as tall as the steeple-tops striding over the face of the young earth. But how much simpler is Nature than man's imagination! Not on the mountain-top did life commence; for the crust of the earth was thin, and unable to bear the weight of the mountains: they were then unborn. Life did not commence on the land; for the land-surface of the globe was a wilderness of bare and heated rock; and life upon it was an impossibility. The first fossils we find give us reason to believe that life commenced in very simple forms, and in the ocean; that the first living inhabitants drew their nourishment from her ample bosom.

To understand what the first animals were like whose remains are found in the rocks, it will be necessary to call in the aid of zoölogy, or the science which treats of animals. Most zoölogists place all the animals of the world in four grand divisions, or, as they are sometimes called, sub-kingdoms.

These are RADIATA, MOLLUSCA, ARTICULATA, and VERTEBRATA. First, the *radiata*, or ray-like animals. These are generally circular, having all the parts of the body disposed in a radiated form, which often gives them the shape of a flower. The parts generally consist of five. The star-fish have five fingers; the crinoids, five sides, five arms, and their fingers are ten, twenty, or some other multiple of five. This division includes the sponges, corals, star-fishes, and all those animals known as zoöphytes, or plant-animals; because, although animals, they much resemble plants in appearance;

being destitute — as is the case, probably, with all the radiata — of head, eyes, or at least what we call eyes, and organs of sensation generally, except feeling; though some, like the sponge, do not even manifest this. They seem to be little more than animated stomachs, "that eat and drink; and then — why, simply eat and drink again." (It is a pity that some of the higher vertebrates so much resemble them.) The star-fish and the sea-urchin, or sea-egg, may be taken as representatives of the radiata, which are generally found in the ocean, and in great numbers. The hydra, or fresh-water polyp, one of the few not confined to the ocean, may be cut into a number of pieces: each piece will grow into a perfect animal; and "even a small portion of the skin soon grows into a polyp." One species may be turned inside out like a glove, the skin on the outside performing the office of a stomach inside; the animal just as ready to take his breakfast as before. The present seas contain about ten thousand species of radiates. The simple forms of life belonging to the radiata have played a very important part in the past history of the globe.

The *mollusca*, whose name is derived from *mollis*, the Latin word for "soft," includes those animals which possess soft bodies, are destitute of a bony skeleton, and generally covered with a hard, protecting shell. They are higher in the scale than the radiata, having a distinct nervous centre; while many of them possess a head, eyes, organs of hearing, and perhaps of smell. They have a heart, with arteries and veins, through which their cold, colorless blood circulates.

The mollusks are divided into six classes, four of which are very important to the geologist.

First, The *cephalopoda*, or head-footed mollusks; deriving their name from the muscular arms or tentacles that surround the head, as in the cuttle-fish and the nautilus.

Second, The *gasteropoda*, or belly-footed mollusks, as the garden-snail and sea-snail, creeping on a broad, muscular foot (whence their name), and usually provided with spiral, univalve shells; that is, shells of one valve or piece.

Third, The *brachiopoda*, or arm-footed mollusks; so called because their organs of motion resemble the "arms" of some polyps. Their shells so much resemble antique lamps, that they used to be termed lamp-shells. The hole which admits the wick in a lamp serves in the brachiopod for the passage of a footstalk, by which it attaches itself to objects at the bottom of the sea.

But few of the brachiopodous shells are to be found in our present seas: they were, however, very abundant in the ancient oceans.

Fourth, The *conchifera*, or shell-bearers, as the term implies, includes the oysters, scallops, muscles, and clams, so well known by everybody.

The third grand division of the animal kingdom is the *articulata*. The name is derived from *articulus*, — Latin for "a little joint." It includes the ringed or jointed animals, — worms, crabs and lobsters, beetles and flies.

In the *articulata*, the brain is in the form of a ring encircling the gullet. The body is symmetrical, — one side like the other; and the skeleton is external, and consists generally of horny rings.

The articulates are the most numerous of all animals at the present day, and probably were during many of the geologic periods; but they are so slight, and so easily destroyed, that it is difficult to find perfect specimens in any of the rocks.

The last grand division is the *vertebrata*, or backboned animals; *vertebra* being the Latin word for "backbone." This division includes all animals possessing an internal, jointed skeleton. They possess a brain, protected by a bony case; and are superior to all other animals in the perfection of their organs of sensation. The *vertebrata* are divided into four classes, — fishes, reptiles, birds, and mammals, or animals that give suck to their young, — commonly called beasts.

A zoölogist giving a list of the various kinds of animals from the highest to the lowest would give them in the following order: —

MAN.
MAMMALS BELOW MAN.
BIRDS.
REPTILES.
FISHES.
ARTICULATES.
MOLLUSKS.
RADIATES.

There might be some difference of opinion about the position of the mollusks, whether they should be placed above or below the articulates; and it might be questioned whether there are not animals lower in the scale than the radiates. With these exceptions, all zoölogists (those worthy of the name) who lived before geology was known as a science, as well as those who live now, would represent the animals of the world as they are represented above. It is certainly a remarkable fact, that a list almost if not entirely identical with this had been made out in the rocks long before man began to think upon this subject. The order as we find it in the rocks is, —

RADIATES.
ARTICULATES.
MOLLUSKS.
FISHES.
REPTILES.
BIRDS.
MAMMALS BELOW MAN.
MAN.

There is a similar difficulty in this list in deciding whether mollusks or articulates are found earliest in the rocks; otherwise this is the order which geology has developed as existing, indicating the successive forms of life that have appeared upon the globe.*

* As there has been some dispute on this subject, especially with regard to the geological position of vertebrates, and statements conflicting with those just presented have been made by some who are regarded as authorities on this continent, it may be well to quote some of the recognized authorities on this subject. Sir Charles Lyell, in his Antiquity of Man, 1863, p. 403, says, "No remains of any vertebrate animal have yet been discovered in the lower Silurian strata, rich as these are in invertebrate fossils; nor in the still older primordial zone of Barrande." Professor Owen, the greatest living naturalist and comparative anatomist, says in his Paleontology, 1861, p. 119, "The earliest good evidence which has been obtained of a vertebrate animal in the earth's crust is a spine, of the nature of the dorsal spine of the dog-fish, and a buckler like that of a placo-ganoid fish. Both have been found in the most recent deposits of the Silurian period, in the formation called upper "Ludlow rock." Since then, it has been stated that fish remains are found in the lower Ludlow in Herefordshire; but, lower than this, there is no good evidence of the existence of fish. In America, we have no certain evidence of fish remains in any Silurian bed. Between the Llanberris slates at Bray Head, in Ireland, in which the

When the question is asked, "What were the first life-forms?" we may not be able to hold up a fossil, as I have heard of a lecturer doing, and say, "This, gentlemen, is the first animal that God Almighty ever made;" for we can never know that the earliest form we find is the earliest that ever existed. We can, however, see in the constantly-increasing simplicity of living beings, as well as their decreasing size, as we go backward in time, a clear indication of what it must have been like.

CAMBRIAN AND LAURENTIAN PERIODS.

AGE OF RADIATES.

The earliest fossil forms, or those found in the lowest or oldest rocks up to this time, are radiates. The oldest rocks in England, Wales, and Ireland, which contain fossils, have been called the Cambrian formation, from Cambria, the ancient name of Wales. It consists princi-

oldest known fossils are found, and the "lower Ludlow," in which are the earliest fish remains, are beds crowded with the remains of radiates, mollusks, and articulates, no less than twenty-four thousand feet in thickness, and representing, therefore, millions of years. So far are all the great types from appearing "simultaneously," as has been affirmed. If fish should ever be discovered in the lower Silurian, as they probably will, still there will be in Europe the immense thickness of the Cambrian beds, with their invertebrate fossils below them; and in America the twenty thousand feet of the Laurentian group, in which radiate fossils alone are found.

Even Agassiz, who has so frequently affirmed that "vertebrates have always existed side by side with radiates, mollusks, and articulates," and whose example has led others to make a similar statement, says in his Essay on Classification, p. 43, that, since the four great branches of the animal kingdom are everywhere found together now, he considers it a "sufficient reason to believe that fishes will be found in those few fossiliferous beds *in which thus far they have not been found.*"

pally of beds of sandstone and slate, and is divided into the lower and the upper Cambrian. In the lower Cambrian are, first, the Llanberris slates, which have a thickness variously estimated at from one to six thousand feet; above these, the Harlech Grits, five hundred feet. The lower Cambrian of Cumberland, England, is estimated by Professor Sedgwick at six thousand feet. In the lower part of the lower Cambrian, at Bray Head, in Ireland, four species of a zoöphyte have been discovered, to which the name of *Oldhamia* has been given, after its discoverer. This ancient radiate form is regarded at the present time as the oldest European fossil; and its simplicity corresponds with what we might reasonably expect from the earliest life-forms, — solid enough to leave an impress on the rock. "One species," says Owen, "presents an axis with radiating groups of branches diverging alternately at regular intervals from either side. The original flexibility of the compound organism is shown by the confused and compressed state in which the whole mass is sometimes found, and from the more or less folded state of the little fans." Dr. Kinahan says, "They have regularity of form; abundant, but not universal occurrence in beds; and are permanent in character, — even when the beds are at a distance from each other, and dissimilar in chemical and physical character."

The upper Cambrian consists of the Lingula Flags, about two thousand feet thick; deriving its name from the lingula, a small bivalve shell, — that being the most common fossil in it; though it also contains radiate forms, and some small crustaceans. Above these are beds, principally of slate, estimated at eight thousand feet, containing but few fossils, — zoöphytes and crustaceans.

There is a formation in America, generally believed to

be older than the Cambrian: it is known by the name of the Laurentian formation (from the River St. Lawrence). It consists of beds of gneiss, quartzite and crystalline limestones; and is said to be twenty thousand feet in thickness. It is found along the north side of the St. Lawrence, and stretches to Labrador on the east, and to Lake Huron on the west. These rocks have been called by some the oldest on the globe, and by others the oldest on this continent; but, till the earth has been more fully examined by geologists, such statements cannot be relied on. There is no doubt, however, of their great age,—far greater than that of rocks, generally supposed, previous to recent discoveries, to be the oldest fossiliferous rocks on the globe.

These rocks had been long suspected to contain fossils, and bodies had been found in them which very much resemble fossil corals; but they had been so much altered by heat, that it was difficult to speak positively with regard to them. Professor Dawson of Montreal has, however, found in these rocks what he regards as shells of *rhizopods*. The name is derived from the Greek *rhiza*, "root;" *pous*, "foot:" they are root-footed animals, whose shells are full of holes, through which pass a number of filaments, looking like roots, by which they lay hold of bodies, and so pull themselves along. These animals are extremely simple in their organization, being destitute of a mouth and stomach: they seem to be as near the lowest round of the ladder of life as we can imagine an animal to be. If these should prove to be fossils, as they are now generally regarded, they take us nearest to that wonderful morn when life first appeared upon the globe.

In this age of radiates I include rocks in which

mollusks and articulates are found; for radiates seem at the time of their deposition to have been most abundant and most diversified. Thus in the upper Cambrian are cystideans, crinoids, corals, and graptolites, — all radiate forms, and representing the three classes into which they are divided. It is probable that unnumbered ages passed away previous to the deposition of these beds, during which radiates were the only existing organisms; but the forms were too slight to leave a fossil impress, or they have subsequently been obliterated by the heat to which they have been subjected. Life marched but slowly in the early fossiliferous periods; and it must have been long, long, after its first appearance, before the vertebrate type was possible.

SILURIAN PERIOD.

AGE OF SHELLS, OR MOLLUSKS.

Above the Laurentian and Cambrian beds we find a succession of limestones, shales, and sandstones, known by the name of the Silurian formation, so called from the Silures, a tribe of ancient Britons who lived in the west of England, where the rocks of this formation lie on the surface, where their peculiar fossils are abundant, and where this formation was first studied.

Both in England and America, the Silurian formation has been divided into lower Silurian and upper Silurian, certain fossils characterizing these divisions in both countries; that is, certain fossil forms are found in them that are found nowhere else, and which enable the geologist to identify the rocks containing them at sight. In the United States, and especially in the State of New

York, where the various beds of this formation are widely exposed and their fossils very numerous, many distinct groups of rocks have been made out, each having some fossils that distinguish it.

The following groups of rocks constitute the Silurian formation as it is developed in the State of New York, commencing at the top: —

Upper Silurian.

LOWER HELDERBERG LIMESTONE.
ONONDAGA SALT GROUP.
NIAGARA GROUP.
CLINTON GROUP.
MEDINA SANDSTONE.
ONEIDA CONGLOMERATE.

Lower Silurian.

HUDSON-RIVER GROUP.
TRENTON LIMESTONE.
BLACK-RIVER LIMESTONE.
BIRD'S-EYE LIMESTONE.
CHAZY LIMESTONE.
CALCIFEROUS SANDSTONE.
POTSDAM SANDSTONE.

Below these, in British America, are

HURONIAN,
LAURENTIAN,

which are included by some geologists in the Silurian.

The lowest of these groups, the Potsdam sandstone, derives its name from Potsdam in St. Lawrence County, New York; where it is about sixty feet thick. It occurs on the south shore of Lake Superior, in Texas, and in Wisconsin; where it is said to attain a thickness of seven hundred feet. There is no doubt that it underlies a large portion of the continent, having been covered up by more recent deposits. The number of fossils found in it is not large: this may be partly owing to the substance of which it is composed; sand being an unfavorable material for the preservation of organic forms. One of the most characteristic is a small bivalve called *Lingula prima* (*lingula* meaning "little tongue;" and *prima*, "first"); the first part of the name describing the shape of the shell, and the other its supposed position in time.

In Wisconsin, many forms of *trilobites* have been discovered in it; some slabs being covered with the casts of them. Fig. 2 represents one from the Potsdam sandstone, Mazomania, Wis.

Fig. 2.

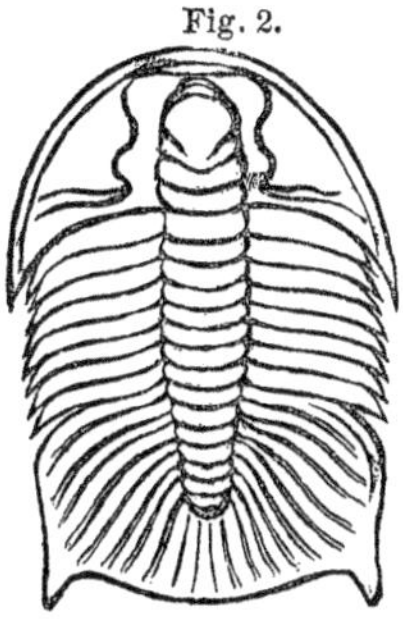

Dikelocephalus Minnesotensis.

The trilobite was an articulate animal covered with a thin crust, or shell; and varied in size, from that of a small beetle to a large crab or lobster. The head, which was crescent-shaped, was furnished with two eyes, composed of a number of lenses like those of a dragon-fly, which, in many species, have been beautifully preserved. The body was composed of many small plates, folded over each other as they are in the tail of a lobster, and divided lengthwise by two furrows into three lobes, from which the name of *trilobite* was given to it. They appear to have swarmed in the early oceans as the shrimps, prawns, and crabs do now, or as the king-crabs on the New-Jersey coast, — animals that more nearly resemble the trilobite than any others found on our Atlantic shore. Hundreds of species have been discovered; and hundreds more, doubtless, remain to be discovered.

Most of the fossil forms found in the Potsdam sandstone are small; but, in the geological rooms at Montreal, tracks of what appear to have been large crustaceans of some kind may be seen. Some of these tracks are in parallel lines, and not less than seven inches wide.

It is not uncommon to find upon slabs of this sandstone wavy appearances, that plainly indicate ripple-marks made on the shore of a shallow sea. Several species of fucoids, or seaweeds, have been discovered in it in New York and in Canada. These leafless marine plants are the oldest types of vegetation known to have existed.

The vegetable kingdom has been divided into six classes. The first includes the seaweeds, the lowest of vegetables; the second, the mosses and liverworts; the third, the horse-tails, the ferns, and the club-mosses; the fourth, the cycads and the firs; the fifth, the pond-weeds, the palms, and the lily tribe; and the sixth, the birch, walnut, sycamore, and all plants having seeds with two lobes.

The first plants that make their appearance belong to the first class. Seaweeds alone of the vegetable kingdom are found below the highest beds of the Silurian formation. Above them we find plants of the second class; above these, plants of the third class, which include nearly all the plants of the carboniferous period. Above the carboniferous, cycads and conifers become abundant; then the palms make their appearance; and, lastly, the plants of the sixth class, which constitute the vegetation of the existing forests in the temperate zones

Above the Potsdam sandstone is the calciferous sandstone, so called because its beds of sandstone contain lime; probably owing to the shells embedded in them. It generally consists of thin layers, and is found both south and north of the Potsdam region in the State of New York, and in Canada, where it extends from Brockville to Ottawa, and sometimes attains a thickness of three hundred feet. It may be seen in Herkimer County, New York, its cavities containing very beautiful and perfect crystals of quartz; and on the Upper Mississippi, where it forms the lower magnesian limestone. In Missouri and Arkansas, it is a lead-bearing rock; and will, I have no doubt, be found, when deeper explorations are undertaken, to be highly so in the great lead-region of the North-west. Occasionally, anthracite is found in it, supposed by Professor Hall to be derived from the marine plants of the time; but probably formed from petroleum altered by the heat, to which the rock has subsequently been subjected. I have seen it frequently in rocks of this age in Canada.

The calciferous sandstone contains many more fossils than the Potsdam, — especially the upper part of it. The earth, by constant radiation, by the abstracting power of the waters, that covered nearly its whole surface, and by its numerous volcanic mouths, was parting with its heat continually; and conditions for life-existence were therefore constantly improving. Remains of sea-plants are common in many localities: they probably luxuriated in thermal waters that were too hot for animals, though these were by no means rare. Mollusks are the most numerous of those preserved, and nearly all of these are gasteropods. A few cephalopods have been discovered; but I shall treat of them when we come to where they were larger and more abundant.

In European beds of this age, fossils that are similar, though not identical, are found; so that distinct provinces of life existed even at that early time.

Above the calciferous sandstone, in the State of New York, we find the Chazy limestone, — named from the town of Chazy, in Clinton County, New York, where it is found, — the bird's-eye limestone, the Black-river limestone, and the Trenton limestone. In the State of New York, these are all distinct, and contain characteristic fossils; but in no other state or country have they all been found. They may all be classed under the head of Trenton limestone, one of the most widespread and highly fossiliferous groups of rocks found in the United States. It receives its name from Trenton, in Central New York, where the West-Canada Creek falls over rocks belonging to this group. In New York, it is in some places three hundred feet thick, but swells in the neighborhood of Ottawa, Can., to the thickness of eight hundred feet. There is a wide exposure of it in Central Tennessee, where it is about three hundred feet thick; and in the lead-region of the Upper Mississippi, most of the lead of that district being obtained from it. The upper part of the Trenton limestone in the lead-region is styled the Galena limestone; having a thickness at Galena of about two hundred and fifty feet. This rock contains a considerable amount of magnesia.

The limestones of the Trenton series abound with fossils. On some of the slabs, as those laid down for flagging in the streets of Ottawa, branching seaweeds are spread over their surface; and in and around Cincinnati, shells, corals, and fragments of crinoids and trilobites, may be seen in the greatest profusion.

The crinoid, of which many species flourished during

the Trenton period, was a radiate with a root like a plant, by which it was firmly attached to the rock. From the root, a long, many-jointed stem arose, and supported a cup-like body containing the stomach: around it were arms, with feathery fingers attached, which spread out and closed at the will of the animal, forming a kind of net, in which the small animals were entrapped that constituted its prey. From the resemblance of some species, when closed, to the bud of a lily, it received the name of *crinoid*, from the Greek *krinon*, "a lily." Fig. 3 represents a group of crinoids as they

Fig. 3.

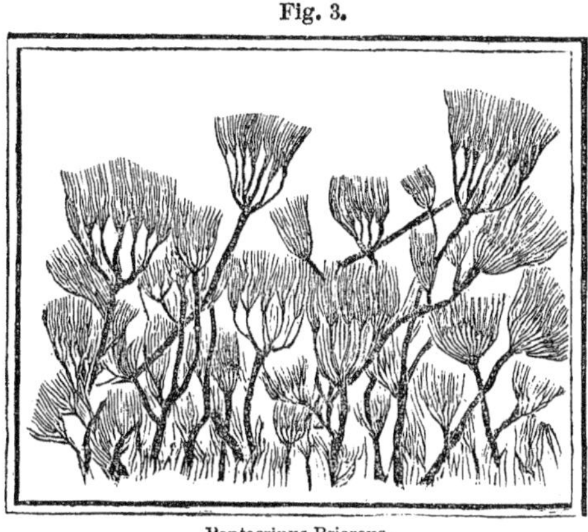

Pentacrinus Briareus.

grew at the sea-bottom, though belonging to a species that flourished at a more recent period.

Mollusks were unusually abundant, many hundred different forms having been discovered; among them a variety of cephalopods, now very rare, but then extremely abundant, and frequently of enormous size. Cephalopods are the largest and most advanced of mollusks in organization. They have a large head armed with strong jaws like a parrot's beak, large and prominent eyes placed on the side of the head, and a large and fleshy tongue. The arms, or feet (for they answer both purposes), are generally from eight to ten, and provided with rows of suckers, which hold so firmly, that it is easier to tear the limb in pieces than to detach it. "If they once touch their prey, it is enough. Neither swiftness nor strength can avail. The shell of the lobster and crab is a vain protection; and even animals many times their size have been soon disabled in their powerful and pertinacious grasp."

One of the most common cephalopods of this time was the *orthoceratite.* Its name means literally straight-horn, from *orthos,* Greek for "straight," and *keras,* "horn;" and a look at the fossil shows at once the propriety of the name. It was a long, straight, chambered shell; tapering from the base, where the largest chamber is, to a point. As many as seventy chambers have been counted in one shell: they were filled with air, and served as floats. Through all these chambers passed a tube called the siphuncle, tapering like the shell: it thus extended from the living animal to the outermost chamber. The specific gravity of the shell with the enclosed animal was less than that of the water: hence it could only sink by an effort. It possessed a funnel, by which it ejected water with great force, and thus reached the sea-bottom, where it could hold by its long, muscular arms. When it wished to reach the surface, all that was necessary was to detach itself; and the lightness of the shell brought it at once to the top. The living nautilus of the Indian Ocean enables us best to understand these extinct forms; for, though the nautilus is coiled, its chambers are similar, the animal passing from a smaller to a larger one as its increased size demands an increase in the size of its tenement. Professor Owen found the crop of one filled with fragments of a small crab. He gives the following passage from an old Dutch naturalist, which I think sheds light upon the habits of the orthoceratite: "When the nautilus floats on the water, he puts out his head and all his tentacles, and spreads them upon the water, with the poop of the shell above water, and with his head and tentacles upon the ground, making a tolerably quick progress. He keeps himself chiefly upon the ground, creeping also, sometimes, into the nets of the fishermen; but after a storm, as the weather becomes calm, they are seen in troops, floating on the water, being driven up by the agitation of the waves. This sailing, however, is not of long continuance; for, having taken in all their tentacles, they upset their boat, and so return to the bottom."

The Trenton period was that in which the orthoceratites attained their greatest size. I have seen one, obtained from the Trenton limestone, in Galena, Ill., eleven inches in diameter at the large end: it was probably twenty feet in length. The accompanying engraving

Fig. 4.

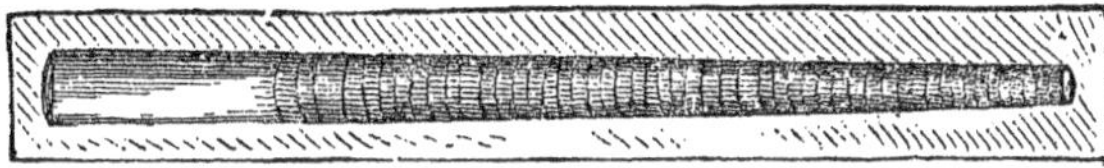

Orthoceras Titan.

(Fig. 4) represents an orthoceratite from the Trenton limestone, Lowville, N.Y.; and, though but a fragment, the original is nearly nine feet long, and the large chamber in which the animal lived is over eight inches in its largest diameter. What enormous tenants must have inhabited these gigantic shells! — their long, muscular arms spread out to seize their prey as they ranged from the surface to the twilight depths of the tepid ocean in which they swarmed. The orthoceratites were carnivorous, and seem to have been to the Silurian seas what the sharks are to more recent ones. Thus early, at least, though doubtless long before, the devourers were busy. Man did not commence the work of killing, as some have supposed; for, ages before he had a being, the existence of myriads depended upon the destruction of others, and the seas were dyed with the blood of the victims. Every foot of ground on which we tread has been a battle-field; and more blood has been shed than there is water in the ocean. It is well that it should be so. Did no animals exist but those that feed upon vegetables, as much life would be possible as the vegetation of the globe would sustain. The animals had all died, then, of disease or old age: now, these all live, all enjoy existence, and perish, without dreaming of the fate that awaits them; but, beside them, an infinite number that feed upon them, and enjoy existence also, — their enjoyment being so much more than could have been possible in any other way.

It is difficult for those unacquainted with geological investigations to conceive of the multitudes of fossils that are found in the limestones of the Trenton group. As a general thing, in Cincinnati, O., Ottawa, Can., and Nashville, Tenn., where the Trenton limestone is found, you cannot pick up the smallest stone without discovering, on examination, that it is a sepulchre containing a multitude of fossil tenants, — shells crowded into shells; trilobites, corals, and shells broken into fragments, and these cemented together by pressure into solid rock; and, of the rock so formed, houses, churches, and cities built: so that, could the animals return and claim what was once theirs, away would fly, like a magician's palace, many of our houses, churches, and gorgeous cities, and an abyss be opened beneath them deeper than the Lisbon earthquake made. Well did Shelley say, "The dust we tread upon was once alive." In many parts of the country this is a sober truth.

Fig. 5.

Calymene Blumenbachii.

Trilobites appear in this limestone in great numbers, and more than twenty species have been recognized. *Calymene Blumenbachii* (Fig. 5) is from Dudley, England. It is one of the species having a wide range; being found, in the United States, from the base of the Trenton limestone to the top of the Hudson-river group.

No feet of trilobites have yet been discovered; and, since they have been found in most perfect preservation, it seems probable, that, if they had possessed feet, we should before this have found them. It is probable that the caudal shield, or tail-portion, which seems to have been movable in all of them, furnished them, acting as a paddle, with the means of locomotion. Thus through the early seas they paddled their light canoes, and set a worthy example to some idle people of the present time.

Corals are very numerous, both branching and massive. About forty species have been described; but many yet remain to be described. A portion of a mass weighing fifteen hundred pounds, from the Black-river limestone, which I include in the Trenton, lying immediately beneath it, is now in the New-York State Collection at Albany. "There is no reason," says Hall, "why extensive coral-reefs may not have margined our early shoals or islands of granite, as those of modern origin do the islands and shoals of our present seas." I have no doubt of the correctness of this view: the masses of coral that I have seen in the Trenton limestone of Central Tennessee and other localities bear evidence to the truth of it.

Still ascending, we arrive at the Hudson-river group,— an enormous mass of shales, slates, sandstones, and, in some places, limestones. Deposits in Anticosti Island, one thousand feet in thickness, belong to this group. It is exposed through the Mohawk Valley and on the Hudson River, from which it takes its name. Towards the West, the beds contain more lime, and fossils are more abundant. They resemble those of the Trenton period in the West, so that the two can hardly be separated. Fig. 6 represents a brachiopod from the Hudson-river group, Iron Ridge, Wis. It is found abundantly near Cincinnati, Richmond, Ind., and many other localities.

Fig. 6.

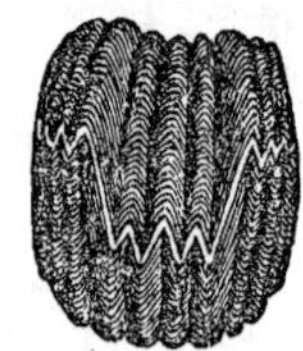

Rhynchonella increbescens.

The most remarkable fossils of this period are *grapto-*

Fig. 7.

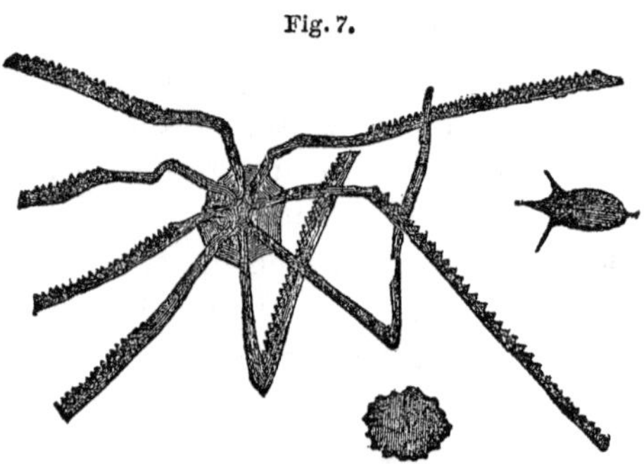

Graptolithus octobrachiatus.

lites (Fig. 7), of which there is a large variety. The

name of the one figured signifies eight-armed graptolite: it was found near Quebec. They are very seldom found as perfect as this, and in their imperfect state gave rise to various conjectures of their nature. At one time, they were considered vegetables; at another, slender orthoceratites: but now they are generally regarded as radiate forms belonging to the acalephs, or sea-nettles. The two small figures represent the young of two distinct species. The shales of the Hudson period abound with the remains of these singular animals, which are more rare in the Devonian, and become extinct in the carboniferous; no remains of them having been found above the coal-measures. Beds in England of about the same age as the Hudson-river group also abound with graptolites, which, since they are generally found in shales, seem to have thriven best in a muddy ocean, as the corals throve best in a clear one.

With the Hudson period closed the lower Silurian. All the Silurian rocks above this are termed upper Silurian.

The Rocky-Mountain range, long and narrow, existed at this time, and probably long before; but between that and land on the east lay a wide ocean, at the bottom of which beds of immense thickness had to be laid as a foundation for the interior of the continent. Toward the south, the land was extending; and a considerable part of Northern New York and Western Vermont was out of the water. A large tract of high land existed in the Atlantic Ocean, the sediment from which made many of the rocks belonging to this period, that are found on the eastern portion of this continent.

Mountain-ranges were heaved again and again, slowly sinking in consequence of the thinness of the earth's crust; rocks were overturned, for on their upturned edges we find in many places that the upper Silurian beds are laid; immense crevices opened at the sea-bottom, and closed after vomiting fiery torrents; and life over wide oceanic areas was frequently destroyed and renewed. By unsteady uplift, the land so increased in the north-eastern part of the United States and South-eastern Canada, that, by the close of the lower Silurian period, a small continent had been formed, over which the tenantless waters ran, sweeping calcareous sediment from the newly-risen land into the undivided sea.

The earth, though much improved, was yet rude and unfinished. The continents were not; and only large islands of naked rock, wave-washed and rain-worn, that sank and rose obedient to disturbing forces within, indicated the places where they should be.

The atmosphere was still heavy with impurities, and hugged closely islands and sea. The night was probably utter blackness, and the day but a diffusion of light.

The gray dawn appears in the east, and gradually extends itself over the watery waste. One by one, rocky islands loom up, faintly visible above the dashing waves. Seaweeds of varied form and hue, their long, leafless arms interwoven, rise and fall with the passing waves. Trilobites by millions, like water-beetles, are sculling their tiny boats in every direction, in pursuit of the soft-bodied animals that swarm by myriads in these more than tropical waters. Cephalopods, larger than a man, are floating on the waters, or diving into its depths; their long arms extended to catch their prey. Meadows of crinoids adorn the sea-bottom; their lily-like bodies, with feathery fringes, more beautiful than tulip-cups, while their many-jointed stems bend to the moving waters like grass in the summer breeze. Here and there, coral-reefs stretch away in waves of beauty; their builders busy in purifying these mineral waters, and preparing them for higher beings than the earth has yet beheld.

We walk along the beach, and find rows of shells and broken corals, heaped as the grass that has fallen before the scythe of the husbandman; every tide bearing new harvests to the shore. We turn to the land, but behold only barren rocks, from whose heated surface vapor is rising in continual clouds. Nothing is heard save the volcano's roar and the dash of the angry waves.

The thousands of feet of rock that we know were deposited during this time within the boundary of what is now the United States, we may be sure took a long period of time for their deposition as sediment, especially when we remember that the land-surface of the globe was small, and there were no large rivers.

Many of the great metallic deposits of North America are found in lower Silurian beds. The copper and gold of Canada East, the copper, iron, and lead of Lake Superior, the gold of Nova Scotia, and the lead of Illinois, Wisconsin, Iowa, and Missouri, are all found here. Metallic deposits are generally found in veins, which were once fissures or crevices in the rocks. At a very early period in the world's history, there were no veins; for, when the earth's crust was yielding, there could be no crevices, and consequently no veins: but as it cooled, and contracted by cooling, ridging up into hills and mountains, crevices by millions must have been formed by the rending of the rocks. Many of these closed again, leaving nothing but a line to tell where they existed; others have been filled with clay and stones swept in from the top, and sometimes by melted rock forced up from below; and, in these cases, we have no metallic veins. Of those in which metals are found, some are *gash veins*, which are shallow, and have been formed without much breaking up of the strata with which they are connected; others, and these are the most important, are *true veins*, fissures in the earth, of indefinite length and depth, sometimes extending for several miles, and probably going down to the molten ocean: no mine has yet gone to the bottom of one. The vein is not a mass of ore: the greater part of it is generally occupied by some rock, different from the strata in which the vein is found, and known by the name of *gangue*, or *vein-stone.* Quartz is most commonly the mineral found in this position.

There are four principal ways by which it is supposed these metallic veins have been filled: —

1. *Infiltration.* — Water heated in the earth's interior, and rising in consequence, as heated water invariably does, brings up with it the minerals and metallic ores which it holds in solution, obtained by circulating through deep beds; and, as it cools, deposits them on the sides of the crevice through which it flows, as lime is deposited from calcareous water on the sides of the kettle in which it is boiled. I have seen rock nearly an inch thick taken out of an engine-boiler, deposited from the lime-water that had been boiled in it.

2. *Segregation.* — The metal having been diffused through the surrounding rock at the time of its deposition, it has been segregated from it, and conveyed, probably by electric agency, into the crevices traversing the rock. Most of the gash veins seem to have been filled in this way.

3. *Injection.* — From the internal fluid mass, metals, in a state of fusion, may have been forced through crevices to the surface. Sometimes metals come up with lava in volcanic eruptions. Veins of injection are found in Lake Superior, where copper has come gurgling up in boiling floods, and consolidated in masses weighing hundreds of tons. In California and Nova Scotia, the gold mixed with quartz seems to have been forced up in a similar manner; and, as an evidence of its original fluid condition, we have the thick veins of quartz poor in gold, and the thin ones rich. The thick ones, cooling slowly, allowed the gold to sink back again by virtue of its greater weight; but the thin ones, cooling rapidly, held the precious article in their grasp.

Lastly, *Sublimation.* — When water boils, it passes into the air in the condition of steam, or vapor. All metals, when subjected to sufficient heat, can be reduced in like manner to vapor. Quicksilver can be boiled and dissipated on a common fire, lead and zinc in a furnace; and, by the heat of the compound blow-pipe, gold can be evaporated. The heat in the interior of the earth is, without doubt, great enough to drive all metals into this state; and, when crevices have been made deep enough, the vapor has risen through them, and condensed upon their sides in the upper part where cool, and produced bodies of ore proportioned to the size of the crevice and the amount of vapor. When the heat from below reached rocks containing water, this water near the crevice would be converted into steam, and the steam would modify the result; softening the rock, and transporting quartz and other minerals from one portion of the vein to another, or from the neighboring rock into the vein. Most of the copper in Canada seems to have been produced in this way, and the gold-bearing ores of Colorado. The vapor of iron, coming up, has united with the vapor of sulphur, and produced sulphuret of iron; the vapors of lead and sulphur, uniting, have produced sulphuret of lead or galena; and in a similar way the sulphurets of zinc, silver, &c., have been formed. The lead found in Wisconsin, Illinois, and Iowa, in sheets or in cubes, lining caves, has, I have no doubt, been placed in its present position by sublimation. Driven from below by heat in a state of vapor, it has passed through the underlying porous sandstone to the limestone, in whose cavities it is now found.* Galena placed in the middle of a tube, and highly heated, on having steam passed through it, is sublimed, in the colder part of the tube, in cubes which exactly resemble the ore. Since the original deposition of the ore by sublimation, many of the deposits have doubtless been changed by infiltration and segregation; and similar changes are still going on.

The fact that the lead-region of the North-west is coincident with the region of no-drift seems to indicate the heated condition of the rocks during the drift period, preventing the formation of ice, and its passage from the north; while the condition of the rocks and their embedded fossils demonstrates that the rocks at some period have been intensely heated.

The great reservoirs of metallic wealth are at a great depth in the earth's interior, — especially the heavier metals, whose greater specific gravity has carried them down; and probably rich mines might be opened anywhere, if men could sink to a sufficient depth, and the internal heat of the earth would permit them to work.

* I heard this idea first advanced by Dr. Charles T. Jackson of Boston.

Although we find metallic deposits in lower Silurian rocks, it must not be inferred that all we find in them were formed during that period; many of them having been deposited during much more recent times. Nor would it be correct to suppose that metallic deposits are confined to rocks of Silurian age. In Colorado, the metals are found principally in metamorphic rocks, some in granitic. The lead of England occurs in the mountain limestone; and copper is found, in Germany, in beds younger still.

In some places, it is evident that ore-forming processes are operating at the present time. Mr. Wright of Liverpool, England, says, "I opened a vein that had not been worked for two hundred years, and from which the ore had been well cleaned out. I found that the sides of the vein had been replenished with the carbonate of lead in crystals of an inch in length, which, no practical man can doubt, have been formed since the period when the mine was worked."

M. Trebra, director of the mines in Hanover, observed native silver and vitreous silver-ore coating the wooden supports left in a mine called Dreyweiber, in the district of Marienburgh, which had been under water two hundred years.

Electric currents traverse metallic veins, as we know by experiment; and these are, doubtless, changing the character of the deposits constantly. We may, by calling in their agency, account for the fact that veins are poor in certain portions of their course when traversing certain rocks, but become rich when rocks of a different nature are reached. No one theory can account for all the facts observed in the mines with which we are familiar.

The Hudson-river group is covered in many places

by a bed of conglomerate, made up of coarse sand and rounded pebbles of quartz. It is called the Oneida conglomerate, and is the lowest deposit of the upper Silurian. But a few feet in thickness in some places, it swells to several hundred feet on the Hudson River. With the exception of some obscure fucoidal impressions, it is destitute of fossils, as might be expected from the coarseness of its composition and the agitation of the water in which it was deposited.

The Medina sandstone overlies this, and consists of beds of shale and sandstone, varying in color from a deep red to light gray; this being their color at Lockport, N.Y., where they are quarried, and used extensively for flagging and for building purposes. There are but few fossils in it; the most abundant being fucoids. At the top of the Medina sandstone at Jordan, Can., I have seen slabs of sandstone covered with their branching arms, interwoven in every direction, and as perfectly sculptured on the solid rock as ever were figures on marble slab by the sculptor's cunning hand.

Some species of shells which are found as low as the Trenton limestone have been discovered in this. There have been no such wholesale destructions and creations as some geologists have fancied. One by one, old forms have perished, and new ones have made their appearance, until we have at length an entirely new set of organic forms. Thus Professor Owen, whose opinion on this subject may be taken as that of our best geologists, says, "It is most probable that the extinction of species, prior to man's presence or existence, has been due to ordinary causes, — ordinary in the sense of agreement with the laws of never-ending mutation of the geographical and climatal conditions on the earth's surface." Partial destructions, sometimes over broad areas, have frequently taken place in the past turbulent history of the earth; but, as we become familiar with wider geological fields we find the range of certain species, once supposed to be restricted to very narrow boundaries, much extended.

Above the Medina sandstone occur beds of limestone, sandstone, and shale, known as the Clinton group. They abound in shells, impressions of sea-plants, and tracks of shell-fish and crustaceans that passed over the soft sand, which, on hardening into solid sandstone, retains an enduring impress. One of the most important shells found in them is a large brachiopod called *pentamerus*, one species of which is represented in Fig. 8, which is a cast of the interior, and is from the Clinton Group, Milwaukie, Wis. Pentamerus means five-parted: a look at the figure will show the propriety of the name. It is a very abundant fossil in rocks of this age, both in Europe and America. Twenty species are known. At Yellow Springs and Springfield, in Ohio, and at Delphi, Ind., they are so abundant, that a block can hardly be found in some of the quarries, destitute of their impressions.

Fig. 8.

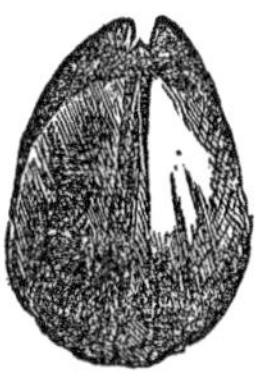

Pentamerus oblongus.

Although these rocks are not more than two hundred feet in thickness in New York, in Pennsylvania they are in some places two thousand feet.

From Nova Scotia to Tennessee a deposit of iron-ore is found in connection with it, which has a thickness, in some places, of thirty or forty feet. This ore is sometimes called lenticular iron-ore, and at others fossil iron-ore, from the great abundance of fossils embedded in it. In Tennessee, where it occurs in beds of great thickness and excellent quality, it is called dye-stone; being used extensively by the inhabitants in dyeing cloth. Situated, as it frequently is there, in the immediate vicinity of coal-beds, it must become of great value in the production of iron.

That solid limestone rock, eighty-five feet thick, over which Niagara's waters pour, and the shale beneath it, eighty feet thick, belong to the Niagara group. As the soft shale wears away, the limestone is left projecting, which makes that covered way, behind the descending sheet of water, along which visitors pass.

This group, consisting of shales and limestones, abounding with fossil remains, is about two hundred feet thick near Niagara Falls, but is said to attain in Pennsylvania a thickness of fifteen hundred feet.

The fossils consist of a great variety of corals; one of the most remarkable and characteristic being a chain coral, in which the ends of the tubes look like the chain stitch. Several species of beautiful crinoids, of which about thirty species are known in this group, many trilobites, and shells in great variety, are found in it.

The locks at Lockport, N.Y., are excavated in the Niagara limestone; and its fossils have been found, and may still be found there, in great abundance. *Dalmania limulurus* (Fig. 9) is one of the most common trilobites, and is frequently found as perfect as here represented.

Fig. 9.

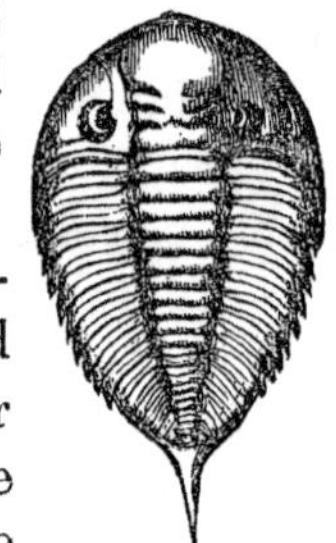

Dalmania limulurus.

This group has been traced by its peculiar fossils over a large part of the United States and British America, — even as far north as the arctic region; indicating the great extent and uniform condition of the ocean in which its life-forms dwelt.

At Lockport, N.Y., there is a thick bed of limestone belonging to this group, which is almost entirely composed of remains of crinoids. It is known there by the name of Lockport granite; the crystals in granite being represented by the crinoidal fragments. The Niagara limestone contains numerous cavities lined with crystals of various minerals, — calcareous spar, strontian, selenite, fluor-spar, and many others. Most of these cavities seem to have been occupied by fossils, which, on decaying, have left the space to be occupied by these minerals, which have been segregated from the surrounding rock. In digging the canal at Lockport, thousands of beautiful specimens of crystallized minerals were taken out of this rock.

The remaining groups of the Silurian formation recognized by the New-York geologists are the Onondaga salt group and the lower Helderberg limestone. The Onondaga group derives its name from the Onondaga Lake, near Syracuse, N.Y., in the vicinity of which its limestones, marls, and shales are found; and from the salt which is found in brine springs and wells bored in them. In the upheaval of the continent, large basins or inland seas appear to have been left filled with sea-water; and these, becoming more salt by evaporation, must have been extremely unfavorable to life, since in the rocks deposited during this time fossils are very rare: indeed, in the true salt-bearing beds, they are almost unknown. The beds of this group are believed to extend through Canada West, and form a part of the Island of Mackinaw.

The salt found in these rocks is never in a solid form, but exists in brine, which is pumped up from depths varying from one hundred and fifty to three hundred and fifty feet. About forty gallons of the water furnish one bushel of salt. Nine million bushels of salt are made from this brine every year in Syracuse and its immediate neighborhood, with no apparent diminution of the supply. The saliferous beds of New York are about a thousand feet thick, showing the continuance of salt lakes or inland seas for long periods; small streams meanwhile carrying down mud, and this becoming thoroughly impregnated by the saline waters.

The Helderberg limestone receives its name from the Helderberg Mountains, in Albany County, New York, where it is two hundred feet thick. It has been found in Pennsylvania, Virginia (where it is eight hundred feet thick), and Tennessee. At the base of this group lie beds of dark, fine-grained limestone, some of which are made into hydraulic cement: hence the name given to the whole, — the water-lime group, — which is well developed at Williamsville, near Buffalo, N.Y. They contain several species of large crustaceans known by the names of *eurypterus* (broad fin) and *pterygotus* (wing-ear), supposed to be allied to the *limulus*, or common horse-shoe crab. The *eurypterus* has a semicircular head, a long, jointed body, and a spiny tail; having one of its pairs of feet flattened into broad blades, or oars (whence its name), probably used for swimming. Similar forms are found in beds of Great Britain belonging to about the same period.

In the lower Helderberg group, about four hundred species of fossils have been recognized, many of them related to species of the Niagara period. Conditions seem to have been pre-eminently favorable for life; as much so as during the Trenton and Niagara periods.

Westward, the rocks of this group thin out, so that

west of Oneida County, in New York, they are hardly known, but, south, are readily recognized by their characteristic fossils in Pennsylvania, Maryland, Virginia, and Tennessee.

It must not be supposed that the groups thus discovered and marked by the American geologists are to be found the world over. While conditions were favorable in one portion of the Silurian ocean for the deposition of sandy sediment, in other portions they were favorable for the deposition of clay; while in others, shells and corals were so abundant, that limestone formed from them was laid down. Conditions thus different, life was necessarily different.

Thus Professor Hall, speaking of the trilobites of the early periods, says, "All those forms requiring calcareous (limy) sediment for their full development will flourish during the deposition of such material, but become diminished or entirely exterminated when a change to argillaceous (clayey) or arenaceous (sandy) deposits takes place: on the other hand, those forms which require a very small proportion of calcareous matter, and flourish in the argillaceous mud, are diminished, or cease altogether, when a calcareous deposition supervenes. The forms which maintain a bare existence, through a series of calcareous deposits, become extensively developed so soon as the nature of the sediment changes; and the same may be said of those requiring calcareous sediment during a period of argillaceous deposits."

So the beds differ much in thickness in different localities. A bed that in one place is five hundred feet thick, fifty miles off may thin out to twenty feet. Where it is five hundred feet thick, sediment may have been rapidly poured in from some river; and, where it is thin, the distance may have been too great for a rapid deposition.

Geologists are not able to run exact lines of division between the various Silurian groups, that shall continue into other continents; for, even at that early period, life-conditions and life-forms, though much more alike than at the present time, were sufficiently diversified to render this impossible.

In the highest of the upper Silurian beds of England and Bohemia have been found a few relics of fishes, and what are supposed to be seeds of club-mosses,—the first traces of land-plants. These early fishes appear to have been small, but extremely carnivorous. They had no true bony skeleton, but a cartilaginous or gristly one in its place; and belonged to the orders Placoid and Ganoid.

All fishes have been divided by Agassiz, according to their scales, into four orders,—placoid, ganoid, ctenoid, and cycloid; a division which, since it is based upon that part of the fish most frequently preserved in a fossil condition, is of considerable use to the geologist.

Placoid is from the Greek *plax*, "a broad plate;" and it includes those fishes which have the skin irregularly covered with plates of enamel,—sometimes large, and sometimes reduced to small points, as in the sharks. Their prepared skins, known by the name of *shagreen*, are familiar to most persons.

Ganoid is derived from the Greek *ganos*, "splendor;" and the fishes belonging to this order are covered regularly with angular scales, composed of horny or bony plates coated with bright enamel. The gar-pike, so common in some of the Western rivers, is a fine specimen of the ganoid: the scales lie in regular rows over the whole body, and overlap one another like the shingles on a house.

Ctenoid, from *cteis*, "a comb," includes all fishes whose scales are toothed at the margin, as in the perch. The *cycloid*, from *kuklos*, "a circle:" the fishes of this order have scales smooth on the margin, and often ornamented, as in the herring and salmon.

The placoids and ganoids flourished for immense periods before the ctenoids and cycloids, which include most of our modern fishes, had any existence.

By the close of the Silurian period, great advance has been made in every direction: the land areas of the globe are greatly enlarged; islands and shallow seas show where continents are destined to be; fishes appear in the ocean, which contains many varied forms of life; humble land-plants clothe the naked rocks; the water is cooler, the air purer, volcanic agency less violent, and the world prepared for higher life-forms.

DEVONIAN PERIOD.

AGE OF FISH.

This period derives its name from Devonshire, a county in England, where the rocks of this age are well represented. In Scotland are immense beds of red sandstone belonging to this age, and known as the Old Red Sandstone, which attains occasionally a thickness of ten thousand feet.

In the United States, the rocks of this period have been divided into several groups, being composed of varied rocks, each containing characteristic fossils which enable the geologist to distinguish them from all others.

The following are the groups of the Devonian rocks, in descending order, as recognized by American geologists:—

Chemung Group.
Portage Group.
Hamilton Group.
Upper Helderberg Group.
Oriskany Sandstone.

There is probably no part of the United States where they can be as well studied as in the State of New York; and most of the groups take their names from places in that State where they are exposed. The smaller groups, unknown out of the State of New York, I shall not notice.

The Oriskany sandstone derives its name from Oriskany Falls, in the State of New York, where it is twenty feet thick. In Pennsylvania, it is about two hundred feet thick, and extends into Maryland and Virginia. There are many fossils in it, especially brachiopods, some conchifers, cephalopods, and trilobites. In some parts of Virginia and Maryland, perfect fossils are found in it in vast numbers. Many of the fossils are casts, and are mistaken for "butterflies," "petrified hickory-nuts," &c.

The upper Helderberg group consists of slates and limestones, principally represented in the West by limestones, which are more than three hundred feet thick in Michigan, but thin out in Iowa to fifty feet.

The upper portion of the group is called the corniferous limestone, which derives its name from the beds and nodules of flint or hornstone which are found in it; *cornu* being the Latin word for "horn." It is usually from thirty to fifty feet thick, and contains numerous fossils, The period of its deposition was an age of crowding crinoids and spreading coral-reefs. The bottoms of the shallow, warm, calm seas, were tenanted by myriads of busy workers; and their products lie spread over immense territories. Professor Hall says, speaking of this time, "With the advent of the fishes at this period, we find that there is a remarkable accession of corals; and the variety of form, and the number and size of species, is much greater than at any preceding period. We may follow the outcrop of this formation, bearing in many places the aspect of a coral-reef, along a line of more than fifteen hundred miles." In the neighborhood of Geneva, N.Y., acres of this limestone lie bare; and we can walk over the surface, and tread under our feet millions of shells, corals, and crinoids of this interesting time. The earliest fish remains yet discovered in America are from the Schoharie grit in the upper Helderberg group.

At the Falls of the Ohio, near Louisville, Ky., the river flows over a corniferous limestone bed, abounding with corals, crinoids, trilobites, and fish bones and scales. Some of these are of large size, indicating gigantic bony-scaled fish in these ancient, coral-bearing seas. The fragments of fish-bones in the white limestone look much like pieces of charcoal. The Devonian period has been termed the age of fish, from the great abundance of their remains that has been found in its beds.

The Rev. John Anderson says, "The remains of these fishes (ganoids) are so very abundant in the yellow sandstone deposit of Dura Den, Scotland, that a space of little more than three square yards, when the writer was present, yielded about a thousand fishes; most of them perfect in their outline, the scales and fins quite entire, and the forms of the creatures often starting freely out of their hard, stony matrix in their complete armature of scale, fin, and bone."

These remains are not as abundant, however, in the Devonian beds of the United States, as they are in those of Great Britain; nor are their remains as well preserved. Many remarkable fishes belonging to this period are now, thanks to the labors of Hugh Miller and others, well known. The *pterichthys*, or winged fish, of which eight species have been described, was a ganoid cased in strong armor, weaker, however, beneath; and in this respect it resembled our iron-clads, and probably for the same reason, — that strength was most needed where the blows of the adversary were most likely to come. "It has less resemblance than any other fossil of the old red sandstone to any thing that now exists,' says Agassiz. Its wing-like appendages were probably used for weapons of defence. This was but a small fish. *Pterichthys Milleri* is little more than an inch long.

Another fish found in the same beds, somewhat resembling the pterichthys, is the *cephalaspis*, or buckler-head, as its name signifies.

Its head resembles a saddler's knife, as Owen observes; and the animal was at first mistaken for a trilobite, some species of which it is somewhat like.

The *coccosteus*, or berry-bone, as its name means, was so called on account of the berry-like projections which ornament its buckler-plates. This ornamentation so much resembles that on the buckler-plates of some tortoises, that it led to the belief that tortoises existed in Devonian times. The jaws and teeth of this fish are particularly interesting; the teeth being chiselled, as it were, out of the solid bone of the jaw, just as the teeth of a saw are cut out of a plate of steel.

The *asterolepis*, or star-scale, was a bony-scaled fish, that sometimes attained a length of twenty feet; so that the Devonian oceans were supplied with fishes of respectable size.

Fig. 10.

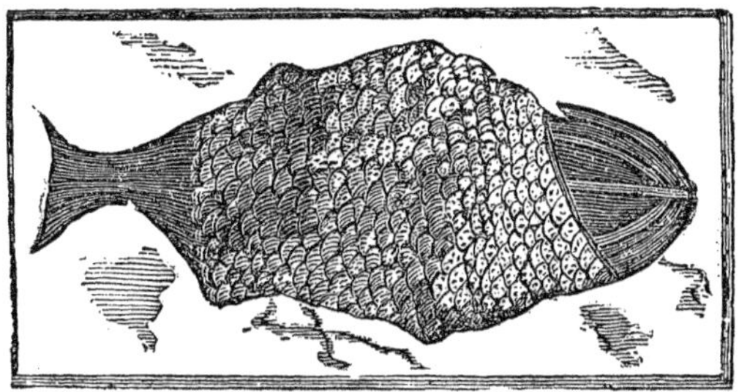

Holoptychius uoblissimus.

Another fish of this age was the *holoptychius* (whole fold). Fig. 10 represents one found in the old red sandstone, at Clashbinnie, Scotland. It lies on its back. "The body" (says Hugh Miller) "measures a foot across by two feet and a half in length, exclusive of the tail, which is wanting; but the armor in which it is cased might have served a crocodile or alligator of five times the size." The jaws are armed with enormous teeth, and the scales are very large and deeply wrinkled.

Returning to a consideration of America, we find above the corniferous limestone the Hamilton group; a series of shales and limestones abounding with fossil remains of shells, corals, crinoids, trilobites, and occasionally of fishes and plants, and generally in an excellent state of preservation: for they were buried in tenacious mud at the sea-bottom, which hardened into shale; and this, when exposed to the atmosphere, returning to mud, the fossils are easily washed out by streams, or may be picked out as perfect as shells on a modern beach.

The Hamilton group extends through New York, a part of Pennsylvania, Ohio, Canada West, and Michigan; and is in some places from eight hundred to one thousand feet in thickness. In some places, I have seen beds of cell coral in Hamilton limestones, the cells of which were filled with petroleum, or rock-oil.

Every one has heard about the remarkable discoveries of oil that have recently been made in the earth's interior; and most persons suppose that it is an entirely new thing in the world's history. Indeed, I have had persons very gravely ask me if I did not suppose that it was an entirely new creation, made expressly, in consideration of the lack of fuel, to supply the world's increasing necessities. This rock-oil is, however, no new thing. We read in the Bible about the rock pouring out rivers of oil; and there is little doubt that the Syrians were familiar with some oil-springs flowing in their day, as one now flows at Zante which was referred to by Herodotus twenty-three hundred years ago. The Egyptians, at a very early period, used asphaltum for the purpose of embalming the bodies of the poor, whose friends could not afford the costly spices by which the bodies of the more wealthy were preserved. This asphaltum indicates the existence of petroleum in the neighborhood; for, when the volatile portions of petroleum pass into the air, the thick, black residuum is asphaltum. The Dead Sea is sometimes called Lake Asphaltites, from the asphaltum found floating on its waters or cast upon its shore. The ancient Babylonians obtained their cement from the fountains of Is (now Hit), on the right bank of the Euphrates, where it still flows, mingled with saline and sulphurous waters. When these oleaginous deposits are developed, the Arabs will see a greater excitement in the Valley of the Euphrates than Layard produced when digging in the ruins of ancient Babylon.

Soon after the discovery of oil near Titusville, in Western Pennsylvania, I saw many pits in that neighborhood that had been dug by the Indians to obtain petroleum, which was used by them to paint their bodies, and sometimes, probably, in religious ceremonies.

At Rangoon, in the Burman Empire, wells have been sunk from two hundred to five hundred feet deep, which yield more than a million barrels annually. Oil has been taken from them for a hundred and fifty years.

This is, then, no new thing; but whence comes it? And in answer to this question we have many theories, some of them sufficiently ludicrous. One suggests, that, since the earth is a huge animal, the rocks its bones, the water circulating in them its blood, the grass and trees its hair, the hills pimples upon its face, and Ætna and Vesuvius eruptive boils, all that is necessary to obtain oil is to bore through the skin into the blubber of the monster, and oil very naturally flows from it. Another supposes, that, during the time of the Flood, the great whales were buried deep under accumulations of mud, in those places where the oil most abounds; and hence petroleum is merely antediluvian whale-oil. It has been suggested, that, since the earth is at some period to be destroyed by fire, the oil was probably prepared against that terrible day when the match will be applied, and the world burned up.

Apart from these ludicrous explanations, however, men of science have considered this question, and rendered their verdict. Professor Silliman says that "petroleum is uniformly regarded as a product of vegetable decomposition." Professor Dana says, "Petroleum is a bituminous liquid resulting from the decomposition of marine or land plants (mainly the latter), and *perhaps*, also, of some non-nitrogenous animal tissues." By many, it is supposed to be a product of coal; and hence the name of "coal-oil," so frequently applied to it. Some suppose that the coal, being subjected to the enormous pressure of the overlying beds, has yielded oil, as a linseed-cake does under a hydraulic press; and I have seen the theory advanced, that the coal, heated (as it evidently has been in the coal-regions of Eastern Pennsylvania), gave off oily vapors, which, rising to the cold region of the upper air, condensed, and subsequently fell in oily showers, making its way as best it could to the hollows of the earth's interior, where the oil-borer finds it to-day.

Facts play sad havoc with these various theories. If the oil comes from coal, it seems strange that it is so rarely met with in a coal-district. I have visited coal-mines in England, Wales, Nova Scotia, Cape Breton, and not less than ten of the United States, but never saw petroleum in a coal-mine, or even smelt it; and this is an article that never waits for an introduction, but salutes the olfactories at once. Of course, if this came from coal, coal-mines would be the places in which to discover it; coal neighborhoods should abound with it, coal-miners be familiar with it; and it should never be found in rocks older than the coal-measures. The contrary of all this is true. Where it is found in the coal-measures, it has been forced up from underlying beds in which it was originally contained.

In this country, nearly all the oil hitherto obtained has been from beds that lie below the coal-measures, and sometimes at a great depth below them. On Oil Creek, in Pennsylvania, it is found by boring in shales and sandstones, sometimes to a depth of a thousand feet; these beds belonging to the Chemung group of the Devonian formation, and many hundred feet below the coal-measures. At Enniskillen, in Canada West, where the oil has at one time come up in springs, and overflowed, leaving a thick bed of asphaltum covering the ground for an acre, the limestone in which borings are made contains characteristic fossils of the Hamilton group of the Devonian formation. The oil-wells in Western Kentucky, and in some parts of Tennessee, are in the Trenton limestone, — that is, in the lower Silurian formation; and I have seen oil even at the base of this. The same oil floats on the surface of a limestone quarry near Chicago, the limestone belonging to the Niagara group of the Silurian formation; showing conclusively that it has no necessary connection with coal.

But may it not have been produced from sea-plants, as coal has been from land-plants, as several eminent geologists have supposed? The quantity of free oil existing in the earth seems to forbid this. I saw a well in Western Virginia which produced twenty-eight thousand barrels in ten months. From three wells near Oil Creek, one thousand barrels spouted in twenty-four hours; and from one, three thousand seven hundred and forty. The "Big Phillips" Well struck oil in October, 1861, at a depth of four hundred and eighty feet. It yielded about three thousand barrels a day. The oil rushed out with such violence, that the well could not be tubed for several days; and it has been calculated that forty thousand barrels of oil were lost in the creek before it could be collected.

The "Noble" Well struck oil in April, 1863. Its daily yield was about fifteen hundred barrels, at which rate it flowed for six months.

There must be lakes of petroleum to render such flows possible. Where are the bodies of fucoids or sea-weeds from which this oil could flow? The sea-weeds of the Silurian and Devonian times (in whose beds the greatest quantity of petroleum is found) were so loose in structure, and contained so little bituminous matter, that their impressions do not even darken the light-colored shales in which they are found embedded. Had these plants been as oily as fish, their bodies would have left dark impressions on the shales, as the bodies of fish do; and if they were not as oily as fish, or as bituminous as land-plants, by what possibility could they produce lakes of oil? If the plants had, indeed, been oily, no oil could have been collected from them, unless preserved from contact with the air and water. Each plant being separated from its companions, on being buried in mud, the oil, supposing any to exist, would have been absorbed by it, and thus lost.

Has the oil been distilled from bituminous shales, as some suppose? I think not. It requires a strong heat to distil oil from shales; and generally, where petroleum is found in the greatest abundance, there is the least appearance of igneous action.

How was it produced, then? It is a *coral*-oil, and not a *coal*-oil. I have in my possession numerous specimens of fossil coral, obtained from Devonian and Silurian rocks belonging to the family of *favosites*, or honeycomb-stone, as the name means, the cells of which very much resemble those of the honeycomb; and, as the cells of the honeycomb are filled with honey, these cells are filled with oil. I have found oil in some specimens nearly as limpid as water; and, by heating the coral, oil runs out readily. I have seen these oil-bearing corals at Smokes Creek, where there are coral-reefs full of it; in the Silurian limestones of Middle Tennessee; at Williamsville, near Buffalo; and in rocks near Penn Yan, in New York. In the State Collection of Fossils at Albany, and in the Montreal Geological Cabinet, there are numerous specimens. Professor Dana informs us, that it flows in drops from a fossil coral at Montmorenci, Can., and at Watertown, N.Y. It might be supposed that this oil filled the cavities of the corals, as it might any other cavity in the rocks: but I have found it repeatedly in these corals, and in no other part of the rock, invariably accompanying the corals, and never connected with any other fossil; these corals frequently in the centre of solid limestone blocks. Reefs of such coral would furnish oil in quantities sufficient to account for the immense deposits that have been discovered. Preserved by them in compact bodies, the oil taking up at least half the space of the coral-reef, we can readily suppose, that when the cells were crushed by the superincumbent weight of rock, or during upheavals and subsidences, cavities and crevices in the earth's interior would be filled by it.

It is, then, an animal production, and not a vegetable one. It is a product of the ocean, and not of the land; being almost invariably associated with salt water from the bottoms of seas that then covered a large portion of

Western New York, Pennsylvania, Virginia, Eastern Ohio, Kentucky, and Tennessee. It is not formed from the bodies of the coral polyps, as some have supposed, — for, when dry, they are a mere film, that could be blown away by a child's breath, — but *secreted* from the impure waters, principally, though not exclusively, of the Devonian times; the coral polyps performing the same office for the water that the carboniferous plants did for the air.

"The old-rock corals," says Professor Owen, "are remarkable for the manner in which they are partitioned off by horizontal 'tabulæ.' Of the one hundred and twenty-nine Silurian corals, one hundred and twenty-one belong to the tabulated divisions." These tabulated corals, that thus flourished in the old times, seem to have been the oil-secreters. And as these tabulated corals diminish in number as we ascend geologically, so, I venture to say, it will be found that the younger formations contain less oil; or, where much oil is found in more recent rocks, it has been forced up from older deposits underlying them.

In some places, petroleum is nearly as transparent as water, as it is found in wells on the Little Beaver, Ohio; at Titusville, of a dark-green color, but quite liquid; at Enniskillen, C. W., thick as tar: and on the ground above, where it has overflowed, you can walk upon it in winter, but find it sticky in summer; as also in Western Colorado, where the Indians told me they had thus caught bears stuck fast in it. On the shores of the Dead Sea it is as hard as cannel-coal, so that the Arabs inscribe sentences upon it; and in New Brunswick at the famous Albert Mine, in Western Virginia, Colorado, and Utah, we find it as a black, solid, hard, shining coal, readily breaking into fragments; while, in the lower Silurian of

Canada East, we occasionally find anthracite coal, as I have seen it in the cells of corals, which is the same article operated upon by heat, — metamorphosed petroleum-coal.

When oil is thus operated upon by the heat of the earth's interior, carburetted hydrogen gas is evolved, which is inflammable: wherever this is found issuing in large quantities from rocks that are below the coal-measures, it indicates the presence of this oil in the vicinity, though it may be at a considerable depth.

The amount of this oil is much greater than is generally supposed. A large part of Canada West, Michigan, Kentucky, Kansas, as well as Ohio, Pennsylvania, and Virginia, are underlaid with it. Western New York contains considerable, as the gas-springs in so many parts of it bear witness. The town of Fredonia, N.Y., has for many years been lighted by the gas issuing from one of these springs; and I have seen a store near Penn Yan heated by the gas, the supply being much greater than was necessary for that purpose.

Fig. 11.

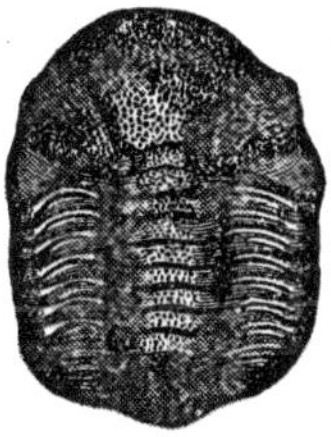

Phacops bufo.

There is a fine exposure of the rocks of the Hamilton group at Eighteen-mile Creek, on Lake Erie; and also on a creek emptying into Seneca Lake, near Bellona, in New York: from which localities countless thousands of shells, corals, crinoids, and trilobites, have been carried off, while the water and the frost are constantly exposing a new crop. Fig. 11 represents a specimen of *Phacops bufo* (lens-eyed toad), one of the trilobites of this period. Its eyes are prominent, and contain sixty-six lenses.

The Portage group, consisting of dark shales and

flagstones, succeeds the Hamilton, and appears to have been deposited in a muddy ocean very unfavorable to life; for fossils are rarely found in it. In some parts of Pennsylvania, it attains a thickness of seventeen hundred feet, but thins out toward the Mississippi.

The Chemung group contains the highest beds of the Devonian formation in America. It is so called from the Chemung River, in the State of New York, where it is well exhibited. It consists of a series of shales and shaly sandstones of an olive or greenish color; and, in many places, abounds with fossils. Large slabs may be taken out completely covered with casts of shells. Impressions of plants are occasionally found in this group, that resemble the ferns of the coal period. It covers a great extent of surface in the southern counties of New York and in North-western Pennsylvania.

In rocks at Elgin, in Scotland, supposed to be of the Devonian age, the remains of a small reptile about six inches in length have been discovered; and, in the same quarry, thirty-four footsteps of a reptile, — probably amphibian. The tracks are about an inch broad; the length of the stride, four inches.

What strange stories this stony record reveals! For millions of years, this ponderous globe rolled round and round; rains fell, winds blew, tides ebbed and flowed; by day the sun, and by night the queenly moon and all the stars, looked down, the guardians of the growing world: yet, during all this immense period, we find nothing higher for which all this was done, nothing to behold all the pomp and glory, but a reptile six inches long! The tree was growing whose fruit should be humanity; and the ages were necessary to knit its giant trunk and perfect its branches.

No remains of reptiles have yet been discovered in American Devonian rocks: they may, however, yet be found. I have no doubt that numerous reptiles lived on this continent, even before the close of the Devonian period; for the first discovery of an animal in a fossil condition is very unlikely indeed to coincide with its first appearance upon the globe. We may as well expect to find the remains of the first man in a fossil condition as to find the first forms of any type.

During the Devonian period, it is evident that the land-surface of the globe was still of limited extent, though constantly enlarging. At its commencement, the northern portion of New York was above the waves; but, at its close, nearly the whole of the State was high and dry, and forms of vegetation covered at least a portion of it. This land was probably connected with a small continent lying to the north and east now covered by a portion of the Atlantic Ocean. The climate was still warm, even to the poles; and the ocean everywhere the abode of forms analogous to those now found only in tropical regions. Devonian fossils have been brought from the shore of the Arctic Ocean, that cannot be distinguished from similar fossils in New York.

As early as the Hamilton period, we know that respectable trees existed. I have seen impressions of their trunks in shales at the Falls of the Ohio, and of long, lance-shaped leaves that must have been of great thickness. It is highly probable that flags and reeds abounded by the water-courses and in the swamps; while low forests of tree-ferns existed where the soil was sufficiently thick for their growth, foreshadowing the age of plants which follows.

Where land vegetation abounds, we naturally look for

insects, which are invariably found accompanying it at the present time. Recently, the remains of flies, some of which were quite large, have been discovered in Devonian slates near St. John, N.B. From their size, we may be sure that these were not the first representatives of the insect world.

LECTURE III.

CARBONIFEROUS PERIOD.

AGE OF PLANTS.

WE have already traced our planet over a considerable portion of the pathway of its progress: at first, probably, a nebulous mass, — a globe of fire-mist, looking like a comet, whose flaming orb startles the drowsy world, as it flies round and round, bathing its burning sides in the frigid space which surrounds it; then a fluid world, the red waves rolling over its agitated surface continually, while the sulphurous clouds hang sullenly above it; then a bare, black, rugged world, whose heated floor is rougher and flintier than the slag of an iron furnace. After interminable ages, it cools, the vapors condense, the valleys are filled, and oceans are born, amid thunderings, bellowings, and battles most dire, — fire and water for ages in deadly combat. Now peace and the spirit of life brood over the world. In the shallow oceans, blood warm, the radiata spread their tentacles, the coral polyps build their stony habitations; while the leathery-leafed algæ take root upon the rocky reef. The mollusks anchor themselves to the rocks, creep with their broad feet over the muddy sea-bottom, or range from the twilight depths to the surface, where they propel their long floats over the waves. The articulata skull their boats in the sheltered bays; while beneath them zoophytes of varied hues bloom in beauty like a garden of flowers. Now moss-like plants carpet the rough rocks upon the land, and ferns fringe them; reeds spring up by the rivers, and tree-ferns wave their heads in triumph on the higher grounds; reptiles crawl and hop on the muddy margin of the sea, the lordliest guests the world has yet entertained. It is still marching on to its goal; and its next step comes up for our consideration.

Above the Devonian formation lies the carboniferous, so called on account of the carbon locked up in its coal-beds, to us the most important members of the formation. Underlying these coal-beds are generally found beds that were deposited at the bottom of the sea, — thick beds of conglomerate or pudding-stone, sandstones, or beds of limestone. In South-western Indiana, in Illinois, Kentucky, Iowa, and Missouri, is a solid limestone, in some places a thousand feet thick: it is known by the name of the sub-carboniferous limestone, from the fact of its underlying the coal-measures, which it invariably does, — a fact worthy of note to those in search of coal in its neighborhood. In England, this limestone is known by the name of mountain limestone, from the fact of its occurrence in Derbyshire, the most mountainous country in England, where it attains a thickness of twelve hundred feet. It has been termed encrinal limestone, three-fourths of it, in many places, being composed of crinoids, or, as they are sometimes called, encrinites. Where the rock has worn away by the action of the atmosphere, the round joints of the crinoids, being harder than the surrounding rock, are frequently found lying loose in the soil, and are gathered by the children in England, and strung for beads. They are called there wheel-whirls, pulley-stones, and St. Cuthbert's beads ; for there is an old legend which attributes their formation to this saint. Thus Walter Scott says, —

> "But fain St. Hilda's nuns would learn
> If on a rock by Lindisferne
> St. Cuthbert sits, and toils to frame
> The sea-born beads that bear his name:
> Such tales had Whitby's fishers told,
> And said they might his shape behold,
> And hear his anvil sound, —
> A deadened clang, a huge, dim form,
> Seen but and heard when gathering storm
> And night were closing round."

Imagine you obtain a glimpse of the saint on the weather-beaten rock, as the storm gathers around him beating away on his anvil, and forging his famous beads. But, as Hugh Miller justly observes, if he made all these beads, he must have been the busiest saint in the calendar ; for they are scattered over the world by countless millions. These fossils bear in Germany the name of *Spangen-steine*, or bead-stones ; *Roeder-steine*, or wheel-stones ; while in Westphalia they are called *Hünen-thranen*, "giant's-tears." I have seen places in Southern Indiana where the body of a wagon might be filled with these fossils in a short time, so numerous are they.

Fig. 12.

Actinocrinus proboscidialis.

Fig. 12 represents the cup, or body, of one of the crinoids from the crinoidal limestone of Burlington, in Iowa. It is called an *actinocrinus* or rayed crinoid, from the ray-like side-arms which project from it. This genus is very abundant in South-western Indiana.

In Illinois, the following groups of rocks, in descending order, have been made out between the base of the coal-measures and the Devonian formation: —

CHESTER GROUP,	500 to 800	feet.
ST. LOUIS GROUP,	50 to 200	"
KEOKUK GROUP,	100 to 150	"
BURLINGTON LIMESTONE,	25 to 200	"
KINDERHOOK GROUP,	100 to 150	"

These are all limestones, and each group contains characteristic fossils by which it can be identified. They abound in fossil shells, and in some places with fish teeth and bones. In Illinois, as also in various parts of Iowa and Missouri, is found a fossil bryozoan, called, from its resemblance to the Archimedean screw, *Archimedes* (Fig. 13); and hence the limestone in which it is found is called Archimedes limestone.

Fig. 13.

Archimedes Wortheni.

The sub-carboniferous limestone has been termed metalliferous limestone, from the abundance of lead that has been found in it in various places. The lead-mines of Derbyshire and Durham in England, and of the Hartz Mountains in Germany, are in this limestone. In Derbyshire and Durham, this lead is found in the vicinity of trap-dikes and volcanic deposits; and the heat and eruptions connected with them may have driven the lead into rocks as high, geologically, as these.

Another name has been given to this limestone, and perhaps the most appropriate of all, — that of cavernal limestone. The great caves of the world are in it, — those of Derbyshire in England, of Indiana, of Missouri, and of Kentucky. The Wyandotte Cave, which has been traced through its various windings for twenty-seven miles, is in this limestone; and the Mammoth Cave, world-famous.

In visiting Wyandotte Cave, in Southern Indiana, you climb a hill, near the top of which the opening to the cave is situated. After sliding down a narrow passage, you arrive at Bat's Lodge, where bats by thousands, hanging down from the roof, show the propriety of its name. After numerous passages and halls have been passed through, you enter the Grand Hall, said to be two acres in extent, with a hill near the centre of it two hundred feet high, composed of blocks large and rude, which have fallen from the roof at various times. On the top of it is a white pillar of stalagmite, formed by continual dropping of limewater upon it: each drop has left a thin film of lime, until this monument has been reared. In some chambers, stalactites, like stony icicles, hang from the roof; while, in others, stalactites growing downward, and stalagmites beneath them, growing upward by the constant dropping, have united, forming alabaster pillars, that seem to support the roof above them. Fantastic forms similarly produced are seen on every side, — tables, thrones, pulpits; while radiant crystals shine resplendently, as you walk, torch in hand, through the echoing aisles.

The Mammoth Cave is by far the largest discovered cave, and the most remarkable. Its long, broad passages, and immense halls, carved out of solid rock, in the mountain's heart, by flowing water; its deep cavities dug by falling cataracts; its wide, deep river; the Workman that accomplished it all in the darkness, where swim the almost transparent blind-fish, — once seen, are daguerrotyped on the soul forever. If you wish to have a sensation of the awful and unearthly, nowhere can you gratify that wish so completely as on Echo River, in the Mammoth Cave. Who but the poet can do justice to such scenes as these? Prentice, after a visit to Mammoth Cave, says, —

"All day, as day is reckoned on the earth,
I've wandered in these dim and awful aisles
Amid the beautiful, the wild, the gloomy, the terrific:
Crystal founts, almost invisible in their serene
And pure transparency; high pillared domes,
With stars and flowers all fretted like the halls
Of Oriental monarchs; rivers dark
And drear, and voiceless as Oblivion's stream,
That flows through Death's dim vale of silence; gulfs
All fathomless, down which the loosened rock
Plunges, until its far-off echoes come
Fainter and fainter, like the dying roll
Of thunders in the distance; Stygian pools,
Whose agitated waves give back a sound
Hollow and dismal, like the sullen roar
In the volcano's depths."

Over a large part of Indiana and Kentucky, there are numerous sink-holes, or swallow-holes as they are sometimes termed, — round hollows produced by water sinking into caves, and sweeping the earth down with it. So numerous are they in some places, that it is impossible to plough the land. A large part of that country is underlaid by immense caves, adorned as never was king's abode, waiting for the eye of intelligence to behold.

While conditions were favorable for the deposition of limestone in some places, they seem to have been equally favorable for the formation of sandstones and conglomerates in others. At Pottsville and its neighborhood, underlying the great anthracite beds, are shales and sandstones of this age, from two to three thousand feet in thickness. In Nova Scotia are red sandstones, and red and green marls, estimated at six thousand feet in thickness. In some parts of Pennsylvania, Ohio, Indiana, Kentucky, and Tennessee, a conglomerate, made of coarse sand and white quartz pebbles firmly cemented together, underlies the coal-measures, and forms an excellent guide in searching for coal. In Northern Ohio and Western New York, this conglomerate is underlaid by blue shale. This having been washed out in places, immense blocks of the conglomerate have been detached, and lie exposed on high points, where they are styled rock-cities. They may be seen south of Olean, in New York, near the Pennsylvania line. Nelson Ledges, on the Western Reserve, and Little Mountain, near Painesville, in Ohio, are other exposures of this remarkable rock, which show how massive strata have been broken up by the action of the elements.

Below the coal-measures, near Pottsville, Penn., Isaac

Fig. 14.

Sauropus primævus.

Lea discovered the footprints of a reptile, which he called *Sauropus primævus*, "first lizard foot." Fig. 14

represents the slab on which they were found, and is one-eighth the natural size. "The animal appears to have had five toes on its fore-feet, and four toes on the hind-pair; longer legs than the crocodile, there being no trace of the dragging of the feet; and a slender tail, which left a groove-like impression. The stride from toe to toe measures thirteen inches; and the feet are about three and a half inches long. The hind-feet stepped upon nearly the same spot as the fore-feet, causing some obliteration of the first impression."

In 1851, Professor H. D. Rogers discovered, in the same red shales, footprints of three species of four-footed animals, all of which were five-toed: the largest had a length of stride of about nine inches. Upon the same slabs that contained these footprints were shrinkage-cracks, such as are made by the sun's heat upon mud; and prints of rain-drops, with signs of water trickling on a wet beach; showing conclusively that the animals which made the prints were out of the water, and must, therefore, have been air-breathers.

Above this come the coal-measures. These are composed of beds of sandstone, shale, coal, and occasionally thin limestones, alternating with each other. The beds passed through in sinking for coal in the Kanawha Valley, in Western Virginia, give a good idea of the way in which they lie. They commenced to dig on a bed of coarse-grained sandstone containing fossil trees, which they found to be eighty feet thick; below that was a bed of iron-ore, a foot thick; then hard, flinty slate, six feet thick; below that was a bed of bituminous coal, four feet thick, with a thin bed of shale beneath it containing impressions of plants; then coarse sandstone, one hundred and fifty feet thick: when they came upon a second bed of coal, four feet thick; below that, sandstone abounding with vegetable impressions, two hundred feet thick; then a third bed of coal, twenty inches thick, with a bed of black shale beneath it, forty feet thick, containing fossil palms; beneath that was a bed of coal, six feet thick, the principal seam of coal in the Kanawha Valley; below that is sandstone, over which the Kanawha River runs.

The number of beds alternating with each other in this way is sometimes very great: the coal generally bears but a small proportion to the accompanying beds. Thus at South Joggins, in Nova Scotia, where the true coal-measures are seven thousand feet in thickness, there are hundreds of beds of sandstones, shales, underclays, and limestones, with seventy-six beds of coal; but there are only forty-five feet of coal in the whole, and very few of the beds are thick enough to pay for working.

What could have produced this singular-looking, black, inflammable rock? How many times this was asked before Science could return an answer! This she now does with confidence. Coal was once growing, vegetable matter. Take up a piece of bituminous coal, and, on closely examining it, you will find in most cases what look like fragments of charcoal; the fibres of the original wood plainly visible in them. By grinding down a piece of bituminous coal very thin, and examining it through a microscope, the very vessels of the wood may be distinctly perceived. Nor is this all: examine the mine where the coal is obtained, and on the surface of the shale, immediately above the coal, you will find innumerable impressions of leaves and branches as perfect as artist ever drew. Dr. Buck land thus eloquently describes the Bohemian coal-mines: "The most elaborate imitations of living foliage upon the painted ceilings of Italian palaces bear no comparison with the beauteous profusion with which the galleries of these instructive coal-mines are overhung. The roof is covered as with a canopy of gorgeous tapestry, enriched with festoons of most graceful foliage, flung in wild, irregular profusion over every portion of its surface. The effect is heightened by the contrast of the coal-black color of these vegetables with the light ground-work of the rock to which they are attached. The spectator feels himself transported, as if by enchantment, into the forests of another world; he beholds trees of forms and characters now unknown upon the surface of the earth, presented to his senses almost in the beauty and vigor of their primeval life."

Still further evidence of the vegetable character of coal is presented in the shale, or underclay, that lies immediately beneath it. In that, we find the very roots of the original plants still ramifying through the shale, once the soil in which they grew. Of the seventy-six beds of coal found at South Joggins, N.S., all but two of them have their accompanying under-clays, with roots of plants in them. This is also the case in Great Britain. In the Welsh coal-basin, every seam of coal is underlaid by a bed of what the miners call under-clay, or fire-clay, in which is found an abundance of stigmaria-roots ramified through it, — a fact first publicly announced by Bakewell, a celebrated English geologist. With the impressions of plants on the shale above, and the roots in the shale below, so universally accompanying the coal, who can doubt the vegetable character of the beds between them?

But how could vegetable matter ever be accumulated in such immense masses as to make beds of coal, such as that anthracite bed at Wilkesbarre, Penn., thirty feet thick? It would require ten or twelve feet of green vegetable matter to make one foot of solid anthracite. Let us transport ourselves to the carboniferous times, and see the condition of the earth, and this may assist us to answer the question. Stand on this rocky eminence, and behold that sea of verdure, whose gigantic waves, as the wind sweeps over it, roll in the greenest of billows to the horizon's verge. That is a carboniferous forest. In the distance, mark that steamy cloud floating above it; an indication of the great evaporation constantly proceeding from it. The scent of the morning air is like that of a green-house: and well it may be; for the land of the globe is a mighty hot-house; the crust of the earth is still thin; its internal heat makes a tropical climate everywhere, unchecked by winter's cold; and it forces plants to a most luxuriant growth.

Descend, and let us wander through this forest, and examine it more minutely. What strange trees! — no oaks, hickories, or elms, no ash, no chestnut, no tree that we ever saw before. It looks as if the plants of a boggy meadow had shot up in a single night to a height of sixty or seventy feet, and we were walking about among the stalks. We are in a gigantic meadow of ferns, reeds, club-mosses, and horse-tails. A million columns rise, whose tufted tops make twilight at mid-day. So close are their trunks together, that we can scarcely edge our way between them. The ground is carpeted with trailing plants completely interwoven; and we can lift up long masses of them, revealing the dark-brown floor on which they rest. What strange trees they are! Here are some covered with scars that look like scales, and give the trees the appearance of gigantic serpents: the scars are the impressions left when the leaves dropped off. Near the tops you can see the leaves standing out on every side from the trunks, which are entirely destitute of branches. Other trees look like Corinthian columns: they are about fifty feet high, and have seal-like impressions in the centre of the flutings, also made by the leaves that once stood out from them all round the tree, as the spokes of a cart-wheel ray from the hub. But have you noticed how soft and spongy the ground is? Shake, and the trees nod for an acre all around. Beneath us lies an accumulation of vegetable matter more than two hundred feet in thickness, — the result of the growth and fall of plants in this swamp for centuries. All things are favorable for the growth of vegetation. The great heat of the ground causes water to rise rapidly in vapor; and this descends in showers, supplying these plants with continual moisture. The air contains a large proportion of carbonic-acid gas, poison to animals, but food to plants, which out of it build up their woody structure. Winds at times level these trees as with a stroke, for their hold on the earth is feeble; and these twining plants weave a shroud over them, and thus the mass increases.

We are now on the edge of a lake abounding with fish: their bony scales glitter in the water as they pursue their prey. It is raining. While in the woods, we were so completely sheltered, that the drops could not reach us. Along the shore are shells that the waves have cast up, and tracks of some large animal. How like the impression of a man's hand some of them are! The hind-feet are evidently much longer than the fore-feet. There is the frog-like animal that made them; but what a size! It must be six feet long. Its head looks like that of a crocodile; for its jaws are furnished with long, strong, sharp, conical teeth, in formidable rows. The shower is over now; and, though the sun does not shine with brilliancy, we can see his place in the heavens. Here is the body of a small, dead reptile. Flies are swarming around it, and beetles feeding upon it; while dragon-flies in immense numbers are flitting around the lake.

The continued growth and deposition of this vegetation during ages must have produced beds somewhat analogous to the peat-deposits of New England and Great Britain. In the Dismal Swamp of Virginia, there is said to be a mass of vegetable matter forty feet in thickness. There is a peat-bog on the banks of the Shannon, in Ireland, three miles broad and fifty long. When conditions were so much more favorable, beds four hundred feet in thickness may have been produced.

This accumulated mass of vegetable matter must be buried, however, before we can have a coal-bed. How was this accomplished? The very weight of it may have caused the crust of the earth to sink, forming a basin, into which rivers, sweeping from the surrounding higher country, carry down mud in their waters which is deposited upon it; its weight pressing upon the vegetable matter, and squeezing it into half its original compass. Sand carried down in a similar manner subsequently, and deposited upon the mud, the mud is pressed into shale; and the vegetable matter, still more reduced in volume by this extra pressure, is better prepared for its final conversion into coal. In time, the basin becomes shallow, from the deposition of sediment on its bottom, and the wearing-down of its outlet; and we have another marsh, with its myriad plants; another accumulation of vegetable matter, by a similar process, buried in like manner. Where thirty or forty seams of coals have been found one below another, we have evidence of land and water changing places as many times.

When vegetable matter is excluded from the air, and under great pressure, it decomposes slowly, parting with carbonic-acid gas; and is first changed into lignite or brown coal, and then into bituminous coal, or the soft coal that burns with smoke and flame. I have been in a coal-mine where the carbonic-acid gas pouring from a crevice in the coal put out a candle placed under it as readily as a stream of water. The high temperature to which the coal has been subjected when buried at great depths has probably assisted in producing the change; and where that temperature has been very high, the coal by the influence of the heat having parted with its inflammable gases, we have the hard or anthracite coal, which burns with little or no flame, and without smoke. It is indeed coal made into coke under tremendous pressure.

The conversion of vegetable matter into coal seems to be going on even now. In Limerick, in the State of Maine, there are peat-bogs of considerable extent, in which a substance exactly resembling cannel-coal has been found. In some of the Irish peat-beds, also, a similar substance has been discovered, having a conchoidal fracture, and hard enough to be worked into snuff-boxes.

It was at first supposed that the plants of the carboniferous times were bamboos, palms, and gigantic cactuses, such as are now found in tropical regions; but a more careful examination of them shows, that with the exception of the tree-fern, now found in the tropics, they differ from all existing trees.

A large proportion of the plants of the coal-measures were ferns; some authorities say one-half. From their great abundance, we would naturally infer the great heat and moisture of the atmosphere at the time when they grew; as ferns at the present time are found in the greatest abundance on small tropical islands, where the temperature is high, and the air is charged with moisture. Ferns are the most abundant plants in the islands of the Indian Archipelago.

Fossil remains of ferns decrease in number continually as we advance from the carboniferous period to the present; and in this fact we have evidence of a gradual decrease of temperature, resulting, doubtless, from the slow cooling of the globe. So high, indeed, was the temperature during the coal period, that the frigid zones were then torrid; for abundance of fossil plants similar to those of our ordinary coal-measures have been found at Melville Island, in latitude 75°, and at Spitzbergen, within the arctic circle.

Next to the ferns, the *sigillaria*, of which about fifty species are known, were perhaps the most numerous plants, and largely contributed to the formation of coal.

The sigillaria, with their tall, fluted columns, attained, when perfect, a diameter of three feet, and a height of at least fifty or sixty feet. These trees have been found erect, or laid horizontally in various coal-mines of Europe and America.

In 1860, four hundred and ninety species of plants had been discovered in the European coal-measures, one hundred and forty-six of which had been found in America;

the whole number of American species being three hundred and fifty. There was a greater resemblance between the vegetation of the two continents at that time; for the species common to both on the discovery of America were few. This resemblance during the coal period was probably owing to the similarity of climate and surface on the two continents, which, we have good reason to believe, was much greater then than now.

The coal formation having been more fully explored than any other, we have become well acquainted with its numerous life-forms. For the first time in the world's history, we have positive proof that life existed in bodies of fresh water; some of the limestones and shales furnishing us with shells, which, from their form, evidently lived in fresh water; while no marine forms are found with them. Land-snails for the first time appear.

Among marine shells, the *productus*, of which Fig. 15 represents a specimen from the carboniferous limestone, Derbyshire, England, is one of the most common. We find the first of this family in the Silurian formation, and they do not pass beyond the Permian. More than sixty species are known: most of them belong to the carboniferous.

Fig. 15.

Productus pyridiformis.

The *spirifer* (Fig. 16) is a shell having internal calcareous spires extending from the centre of the shell outward: from these it receives its name. Many species of this family are found in the Silurian and Devonian. The one represented is from the carboniferous limestone of Kildare, near Dublin, Ireland.

Fig. 16.

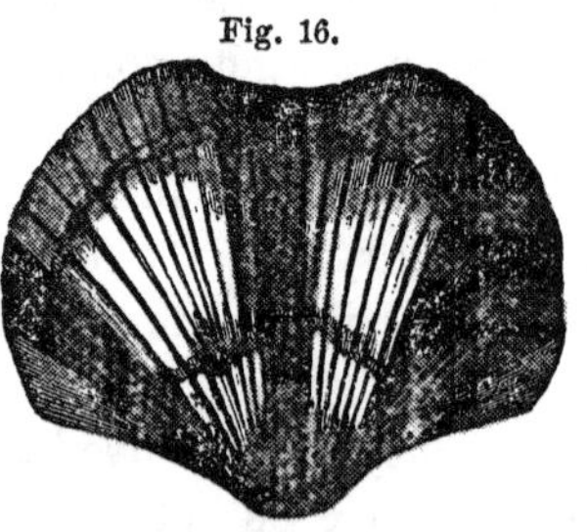

Spirifer pinguis.

Cephalopods are not as common as they were during the two previous periods; but some specimens of the orthoceras, nautilus, and goniatite (Fig. 17), are occasionally met with. The goniatite, from the Greek *gonia*, "an angle," derives its name from the angular division between the chambers. A hundred and fifty species have been described from the Devonian formation. About thirty species of orthoceratites are described from the English carboniferous beds; and no species has yet been discovered in any more recent formation.

Fig. 17.

Goniatites Ixion.

Trilobites were few in number, and approaching utter extinction: it is very rare that a specimen is found in the coal-measures; and above this, it is said, no specimen has yet been discovered. The wings of insects have been found in Germany, England, and the United States; and from them we learn that dragon-flies, ants, locusts, and cockroaches may boast of a very high antiquity. People talk about tracing their ancestry to the Norman Conquest; but what is that, when a gentleman-beetle might trace his to a time when Normandy was not, and the rock had yet to be made of which it was to be formed!

Centipedes, or thousand-legged worms, have been found in Nova Scotia, which had burrowed into a sigillaria trunk.

Fish remains are abundant. I have seen thousands of teeth and bones on sandstone slabs formed at the bottom of some sea then covering Northern Ohio. All the fish of this period, as well as those of the Devonian and upper Silurian, possessed vertebrated tails; that is, the back-bone, or the cartilage which occupied its place, was prolonged into the tail, — a peculiarity which we now see in the bony pike, shark, and sturgeon, living representatives of a very ancient type. In nearly all cases, this made a one-sided tail.

In the younger rocks we find the lop-sided fishes, or those with heterocercal tails, become fewer in number, and the homocercal fishes, or those with equal-lobed tails, take their place ; so that, as Agassiz has observed, " the progress of the ages is marked in the tails of the fishes."

Many large fish belonging to the genus *megalichthys* (from the Greek *megas*, " large," and *ichthus*, "fish") have been found in the carboniferous strata of Edinburgh, Leeds, Manchester, and other localities in Scotland and England. These fish combined many of the characters of a true fish with many and striking resemblances to reptiles. The teeth, more especially, resemble those of crocodiles. One tooth was found measuring nearly four inches in length, and nearly two inches in breadth at the base. The large teeth were accompanied by small ones, alternating with them, and distributed over the whole of the inside of the mouth. The teeth are cone-shaped, and have a conical hollow at the base, in which the next tooth is prepared, so that these necessary implements may never be wanting : the same provision is found in the teeth of reptiles.

The scales of the megalichthys are formed of enamel, having a most brilliant lustre : they are generally angular, but sometimes rounded. Their rounded scales have been found five inches in diameter.

In 1852, Professors Dawson and Lyell discovered in the coal-measures at Joggins, on the Bay of Fundy, the bones of two small, frog-like reptiles, and the shell of a land-snail, " in the interior of an erect tree, mingled with the sand, decayed wood, and fragments of plants which had fallen into it after it became hollow." Since that time, Professor Dawson has discovered another fossil stump, containing the remains of four small reptiles, a centipede, and a land-snail. In all, there were four species of reptiles in these two stumps. How little we know of the numerous forms of life that must have swarmed in those old carboniferous swamps!

In the coal-measures of Bavaria, many fossil remains have been found of an animal styled the *archegosaurus*, or first lizard; its namer supposing it to be the oldest form of reptile. Professor Owen, the highest living authority on these subjects, regards it as a transitional form between the frogs and the ganoidal fishes. "It affords," says he, "the most exemplary instance of a transitional form, on the derivative hypothesis, of an air-breather from a water-breather."

Four new genera of reptiles have been recently found in the coal-beds of the Jarrow Colliery, Kilkenny, Ireland. One of them was a reptile nearly seven feet long. They are *labyrinthodonts*, which were very numerous after this, and attained a gigantic size. I shall refer to them again.

To obtain coal in this country is generally an easy matter, cropping out, as it does, on the banks of so many streams. Many beds of it may be seen on the Pennsylvania Central Railroad, on the western slope of the Alleghanies, on the banks of the Alleghany River, and on the Ohio, as well as on various streams of Ohio and Virginia. Nothing more is necessary in such cases than for the miner to "drive" into the bank, bring out the coal, and drop it into the flatboat moored immediately beneath; while the same level by which the coal is brought out serves to drain the mine of water.

In England, and on the continent of Europe, however, the process for obtaining coal is a much more difficult one; all the coal that could be obtained easily having been long since exhausted. The first thing to be done is to discover where the coal lies; which is no easy matter. For this purpose, boring is necessary, as for oil. A steel drill about four inches in diameter is lifted, dropped, and turned, until a hole of equal size is made down to the coal-bed, whose presence and thickness can thus be ascertained. If a bed two feet thick, it will not pay to work; and down goes the drill to a second, perhaps a hundred feet deeper. This is three feet thick; but that is not thick enough for profitable working; so down to a third. This is six feet thick. Now the work of sinking a shaft commences; for a hole must be made large enough to bring up the black treasure discovered.

The miners dig, therefore, a well-like shaft ten or twelve feet in diameter; but, before going far, water rushes in more rapidly than they can bail it out. A steam-engine is necessary to pump out this water, and to keep the mine clear when it is in operation. Another one is needed to hoist up the rock and coal; and then the work goes on with increased speed. When the coal is reached, and galleries are cut in it to obtain the coal, the miners soon experience a lack of pure air, — especially after blasting with gunpowder. For the purpose of supplying the mine with pure air, a second shaft is generally sunk, sometimes at a distance of half a mile from the first; galleries are made underground from one shaft to the other; and at the bottom of the second a large furnace is erected, and a fire made, which is constantly supplied with coal. This causes a current of air to rise to the surface up this shaft, which is called, on this account, the "up-cast shaft." Down the other shaft, which is called the "down-cast shaft," goes the pure air, regulated by trap-doors in the galleries, sweeping through the mine, carrying off the impurities from every portion to the furnace, and so to the surface.

In Great Britain, there are about twelve thousand miles of coal-lands; and more than eighty million tons of coal are taken out every year. It has been calculated, that, at the present rate of use, the coal of this island will not last more than two hundred years. Already it has become a serious question in England, "What shall we do when the coal is gone?"

In the United States, there are about one hundred and twenty thousand square miles underlaid by known workable coal-beds, besides what remains yet to be discovered. With this immense stock of fuel in our cellar, what a future there will be for the house of Jonathan! Long will it be before the question will be seriously asked in this country, "How long will our coal last?"

Although bituminous coal is generally the product of vegetation, it is not invariably so. In Albert County, New Brunswick, there is a large fissure, in places seventeen feet wide, which is filled with a jet-black, shining coal. It has been worked to a depth of seven hundred and fifty feet, and apparently continues to a much greater depth. This coal is now acknowledged to be solidified petroleum; and is therefore, as I think, an animal product.

All petroleum-coal that I have seen and heard of occupies veins, generally perpendicular, instead of horizontal beds. There is one in Ritchie County, Western Virginia, another in Scotland, several in Cuba, and others which I discovered, partly in Utah, and partly in Colorado, on White River. The coal of these can scarcely be distinguished in any way from the Albertite of New Brunswick. The largest deposit forms a perpendicular vein about three feet wide, which we traced for more than three miles. In the same region is a bed of highly bituminous shale, equal to cannelite, which I found at various points; indicating that it extends over twelve hundred square miles. The bituminous deposits of this country are not all discovered yet. We have sleeping servants under ground that future generations must waken.

THE PERMIAN FORMATION.

In the eastern part of European Russia, and principally in that part of it which constituted the ancient kingdom of Permia, are limy flagstones, magnesian limestones, bituminous shales, and red and green sandstones; some of these beds sufficiently impregnated with copper to pay for working. They cover an extent of country twice as large as France; and, from their occurrence in and around Perm more extensively than in any other explored locality, the formation to which they belong, which is next above the carboniferous, is styled the Permian formation.

This is the highest or most recent paleozoic formation. Paleozoic means, literally, "old life;" and the paleozoic rocks are those of the Silurian, Devonian, Carboniferous, and Permian formations, containing the oldest animal and vegetable forms. The fossils found in the Permian rocks more closely resemble those found in the rocks below, or those that are older than the Permian, than those that are found in the younger formations. M. Pictet shows that fifty-six generic forms out of a hundred pass from the earlier periods to the Permian; and hence the propriety of placing this with the old-life formations. Murchison says, "In ascending above the highest Permian deposits, the geologist takes a sudden and final leave of nearly every thing in nature to which the words 'primary,' 'primeval,' or 'paleozoic,' have been or can be applied."

Caution is necessary, however, in accepting this statement. Thirty-three out of a hundred generic forms belonging to the older periods are found in the trias, the formation immediately above the Permian; and, when all the Permian and triassic beds are known, we shall find, as in all other formations, that life-forms change almost insensibly from the one to the other. Thus he says, in his last anniversary address as President of the London Geological Society, "We have been obliged to give up the theory of great breaks between successive formations, as we find a gradual passage from one geological formation to another, evidenced by a gradual dying-out of the pre-existing forms of life, and the gradual introduction of newer."

Owen also says, "The sum of the animal species at each successive geological period has been distinct and peculiar to such period. Not that the extinction of such forms or species was sudden or simultaneous: the evidences so interpreted have been but local. Over the wider field of life at any given epoch, the change has been gradual, and, as it would seem, obedient to some general, continuously operative, but as yet ill-comprehended law."

In the Permian rocks of Germany there is a large development of magnesian limestone, underlaid by dark bituminous shales highly charged with sulphuret of copper, which is called *Kupfer-schiefer*, or copper-slate. These slates have been worked for many years for the copper that they contain; and fossil fishes have been found in them in great abundance.

The copper-slate of Mansfield has a thickness varying from one and a half to two feet, and is worked in numerous mines by a process called "crooked-stick work," — the miners crawling and working, lying upon their sides, in cavities eighteen or twenty inches high; the roof over them supported by pieces of bent timber or crooked sticks.

The Permian formation is well developed in the north of England, where it consists of magnesian limestones and sandstones, having a thickness of one thousand feet. The harder varieties of the Permian limestones make excellent building-material; and the new Houses of Parliament, in London, are constructed of this limestone, which is thought to be the best building-stone of England. Permian rocks have recently been discovered west of the Mississippi River, principally in Kansas; but we have no knowledge of their precise boundary. They consist of limestones, sandstones, and shales; and their greatest thickness is less than a thousand feet.

The plants of this period belong to genera that are common in the coal-measures, but are either wanting or very rare in the formations above. Silicified stems of ferns have been found in Saxony, and silicified trunks of coniferous or cone-bearing trees in England.

The trilobite has not been found in the Permian.

This family came into existence early, attained its greatest development about the middle of the Silurian period, diminished in size and number in the Devonian, was reduced to a few small species in the carboniferous, and with the Permian seems to have died out to appear no more. The limulus, an animal somewhat resembling the trilobite, seems to have taken its place. It appeared in the coal-measures: large species have been found in the Permian beds of Russia, and some species still exist in our present oceans. The king-crab, or horse-shoe crab, is a familiar example.

Between fifty and sixty species of fish have been found in this formation, some of which are identical with those found in the carboniferous formation. All these fish have heterocercal tails. In one locality in Ireland, a small species is found in such numbers, that more than two hundred have been counted on a slab two feet square. The ichthyolites, or fossil fishes, of the *Kupfer-schiefer* of Germany, may be found in almost every museum of Europe. These specimens, says Mantell, are splendidly invested with copper pyrites, and their scales have the appearance of burnished gold.

Many remains of reptiles have been found in the Permian deposits; the first by Dr. Spener, a physician at Berlin, in 1710. It was obtained from a copper-mine in Germany, at a depth of one hundred feet. Other specimens have been discovered since. They were aquatic reptiles, furnished with sixty-eight sharp, conical teeth; and probably fed upon the fishes which we know flourished in the waters which deposited the copper slates of Germany.

Some remains of reptiles similar to these have been

found near Bristol, in England, and, according to Dr. Emmons, in the Permian strata of the Chatham coal-field, in North Carolina.

"The old must pass away that the new may come in." So pass away, with the Permian formation, old oceans, old islands, old shells, old fishes. As the stars sink, one by one, in the west, and new stars rise in the east, to be succeeded by the dawn, and then the day; so, through the night of the past, sank the old life-forms, and were succeeded by the new, approaching nearer and nearer to the dawn of that day in whose morning we live.

SECONDARY PERIOD.

AGE OF REPTILES.

Trias, or *New Red Sandstone.* — The secondary formation is divided by Lyell into five divisions: —

1. Trias, or New Red Sandstone.
2. Lias.
3. Oölite.
4. Wealden.
5. Chalk, or Cretaceous.

The trias, or new red sandstone, derives its name of trias from the fact, that, in Germany, it consists of three different groups of rocks, called *Bunter Sandstein*, or colored sandstone, *Muschelkalk*, or mussel-chalk, and *Keuper*, which is supposed to be a miner's term, used to designate a group of red and green marls, and shales.

It receives its name of new red sandstone from the fact, that, in England and in this country, it is formed principally of red sandstones; and to distinguish between them and the old red sandstone, belonging to the Devonian formation, it is termed the new red sandstone.

This formation has received the name of saliferous formation, from the great abundance of salt associated with its beds. Salt is not by any means confined to the new red sandstone, but, in Europe, is found more commonly connected with rocks of this age than with any other.

At Droitwich, England, one salt-spring yields a thousand tons of salt a week. In boring for this brine, they pass through four layers of salt, eighty-five feet in aggregate thickness. The brine is so strong, that it yields one-fourth of its weight in salt. In Cheshire, there is a bed of salt eighty-five feet thick, then a bed of marl thirty feet thick, below which is a second bed of salt one hundred and six feet thick.

At Cardona, in Spain, there is a salt-mountain several hundred feet high. At the foot of the mountain is a rivulet, which in rainy seasons swells into a river, and carries down so much salt as to destroy the fish. "Nothing," says Count Laborde, "can be more beautiful than the appearance of this mountain at sunrise. Besides the beautiful form which it presents, it appears to rise above the river like a mountain of precious gems, displaying the varied colors of the rainbow."

The chief repository of salt is in Poland. Mines have been worked ten miles from Cracow to a depth of more than one thousand feet. These mines are entered by eleven shafts, with galleries at five different levels, leading to a labyrinth of passages and excavations, whose total length is two hundred and seventy miles. Chambers have been dug out one hundred and fifty feet high. One was fitted up as a chapel, and dedicated to St. Anthony: it contained an altar, statues, columns, and pulpit, all of salt. Dr. Darwin, in his "Botanic Garden," referring to this mine, says, —

> "Thus caverned round in Cracow's mighty mines,
> With crystal walls a gorgeous city shines;
> Scooped in the briny rock, long streets extend
> Their hoary course, and glittering domes ascend;
> Down the bright steeps, emerging into day,
> Impetuous fountains burst their headlong way,
> O'er milk-white vales in ivory channels spread,
> And, wandering, seek their subterranean bed."

The question very naturally presents itself, Whence came these enormous masses of salt? There was a time when they were not, and a time when they began to be. Some have suggested, that salt being in some places a volcanic product, salt-springs flowing in the neighborhood of many volcanoes, these saline deposits may have been formed at some past time, during a period of great volcanic activity; the salt carried into lake-basins, and there evaporated.

Their origin has also been referred to the evaporation of the water of ordinary salt-lakes, leaving the salt as a sediment. All lakes having rivers flowing into them, and no outlet for their waters, are salt-lakes. All rivers contain more or less salt: this, flowing into a basin, is constantly increasing; for, though the water is evaporated, the salt is, of course, left behind. In time, so much salt is deposited in the lake, that the water cannot hold any more in solution, and it settles to the bottom by its weight. Put a spoonful of salt into a cup of water, the water will dissolve it readily; put in another, the water will dissolve it less readily; by adding still more, the water at last can take up no more salt, and it lies in a solid form at the bottom of the cup. The River Jordan has many salt-springs flowing into it, and is therefore constantly carrying salt into the Dead Sea, which is thirteen hundred feet below the level of the Mediterranean, and consequently has no outlet. On sounding it, Lieut. Lynch brought up crystals of salt from a depth of one hundred and sixteen fathoms, showing that such a process is now going on there. Lake Indersk, in Siberia, has a crust of salt at the bottom, hard as a stone, and perfectly white. There are salines in New Mexico, old lake-beds, five or six miles in circumference, at the bottom of which are beds of salt of unknown depth.

It is said that four gallons of the water of the Great Salt Lake, in Utah, will make one gallon of salt; and if the Bear and Jordan Rivers, which flow into it, should be cut off by any natural convulsion, it would form an immense salt-pan, the water would evaporate, and there would be one-fourth as much salt left at the bottom as there is now depth of water in the lake.

We can readily see, then, how some salt-beds could be produced. But we have facts which point to the ocean as the source of some saline deposits. In the Polish salt-mines, bivalve shells and the claws of crabs are met with in the marls associated with the salt-beds. In Bolivia, sea-shells are found with salt-beds, which are more than six thousand feet above the present sea-level.

The ocean is the great reservoir of salt, and has been, probably, ever since it existed. Estimating its average depth at five thousand feet, there are in its waters thirty thousand million of millions of tons of salt. Doubtless a large proportion of the salt contained in the earth's strata came originally from this grand reservoir. The new red sandstone period, we know, was one of unusual volcanic activity. That mountain-ridge of greenstone in Connecticut and Massachusetts, of which Mount Holyoke and Mount Tom are the highest portions, is believed to be of this age, and indicates, to some extent, the size of the lava-torrents of that time. The Palisades on the Hudson; the trap of the Island of Staffa, in which Fingal's Cave has been hollowed; the trap forming the Giant's Causway, in Ireland; the long bed of trap on the shores of the Bay of Fundy, — all believed to have been formed during this age, — testify how fearfully the Earth's fires raged during this period, and how she poured out her boiling tides.

Imagine an arm of the sea twenty miles long, ten miles broad, and a thousand feet deep, cut off from the main body by an eruption of lava, which flows across it. The heat produced by it causes the water to boil like a kettle on a fire, as the recent lava-flood at Hawaii caused the sea to boil. The water is evaporated, and twenty-seven feet of solid salt are left at the bottom; for in one thousand parts of sea-water there are twenty-seven parts of salt. If, instead of being a thousand feet thick, it had been four or five thousand feet, the salt-bed produced would have been of corresponding thickness.

But, after proceeding so far, the question arises, Where did the sea obtain its saltness? For man is an inquisitive animal, whose questions corner at last the wisest. There was a time when the ocean was not, if we have read aright the story of the earth which the rocks relate. Where was the salt then? Lieut. Maury thinks that the sea was created salt, and suggests, that, had it been otherwise, the present animals living in it could have had no existence; but then, I suppose, forms of life just as perfect could have existed in it, had it been fresh, as we find all large bodies of fresh water tenanted to-day. Sir Richard Blackmore asks, "What does the sea from putrefaction keep?" and then in answer says, —

> "Should it lie stagnant in its ample seat,
> The sun would through it spread destructive heat.
> The wise Contriver, on his end intent,
> Careful this fatal error to prevent,
> And keep the waters from corruption free,
> Mixed them with salt, and seasoned all the sea."

We may excuse the poet where we criticise the man of science; but I should like to ask him why such a large body of water as Lake Superior is fresh, and how it is preserved from putrefaction; and why such a small body of water as the Dead Sea is so salt, so highly seasoned, that it would seem as if the salt-cellar had been dropped into it.

There must have been a time when there was no salt, the elements alone existing of which salt is a combination. The chemist informs us that salt is chloride of sodium, or a union of chlorine (a heavy gas, being two and a half times heavier than the air we breathe) and sodium (a light, bright, beautiful, silvery metal; so light, that it will float on the surface of water). By burning sodium in chlorine gas, the two unite, and salt is the result; or it may be produced by saturating soda, which is the oxide of sodium, with hydrochloric acid, and evaporating to dryness. When, at an early period of the world's history, acids distilled from the clouds, and the earth was a grand chemical laboratory, either directly or indirectly, salt was produced on its surface with many other minerals; and, as the rains fell, all that were soluble were leached into the hollows of the globe, and what remain unprecipitated give to the sea its saltness. The sea is but a great vessel into which the impurities of the land have been drained: and many of our geological beds are but those impurities in a compact form, buried generally from sight; both land and water benefited by the process, — that is, prepared for the use of higher organic forms.

For it is not common salt alone that the sea contains, but many other substances mingled with it. In one hundred parts of solid matter contained in the water of the ocean, there are of

Common salt, or chloride of sodium . . .	78.61
Chloride of potassium	1.34
Chloride of magnesium	8.56
Sulphate of lime, or gypsum	3.47
Sulphate of magnesia, or Epsom salt . . .	6.42
Carbonate of lime	.27
Of residue undetermined	1.33

In this undetermined residue, bromine and iodine are contained, as they are known to exist in sea-water.

To leach these saline impurities out of the land seems to have taken a long time. We have no evidence of the existence of bodies of fresh water on the globe until after the Silurian period. Not till we arrive at the Devonian formation do we discover shells whose form indicates that they were fresh-water inhabitants, that the rains of ages had at last purified the land, and fitted it for the sustentation of life.

With salt, we generally find sulphate of lime, or

gypsum, associated. This is the case in England, at Syracuse in New York, and at Grand Rapids in Michigan; so that when we find gypsum we look for salt, and when we find salt we look for gypsum, in the neighborhood. When gypsum is ground, as it is in mills, like corn, to a fine powder, it is then used as a fertilizer; and is on many lands of great value, as farmers generally know. When this powder is exposed to a heat of about two hundred and thirty degrees, it parts with its water of crystallization (one-fourth of gypsum being water), and becomes a dry powder called plaster of Paris. This, when mixed with water again to about the consistency of cream, returns in a short time to solid stone. It is of the greatest value on this account for making casts; and dentists and artists generally make great use of it. The plaster casts of dogs, monkeys, and men, carried by Italians on their shoulders through our streets, are made of this material; and there are few halls or parlors which it does not assist in decorating.

It is a sulphate of lime, formed by a union of sulphuric acid and lime. Sulphuric acid is one of the products of volcanoes. Streams of it flowing over or through limestone rocks would dissolve the limestone most rapidly, and form sulphate of lime; this, carried into lakes or the ocean, would settle by virtue of its weight to the bottom, and form a bed of gypsum. In other cases, where the ocean or lakes held in their waters much lime, sulphuric acid poured into them would change this lime to gypsum, which would sink, and form a bed. Experiments that chemists now try cn a small scale, Nature tried ages before on the grandest scale, and with the most important results.

The new red sandstone exists, it is supposed, in the

Connecticut Valley, from the north of Massachusetts to Long-Island Sound, a distance of one hundred and ten miles, with an average breadth of twenty miles. Another range extends from Rockland, on the Hudson River, through New Jersey, Pennsylvania, and part of Virginia. It is three hundred and fifty miles long, and from five to thirty miles broad. Several other ranges are known in Virginia and North Carolina. It is evident, that, where these ranges are, valleys once existed parallel to each other, and having their longest direction from north to south. These valleys were occupied by water, either as lakes, estuaries, or arms of the sea; receiving sediment continually as it was poured in from the neighboring hills, either as pebbles to make conglomerates, coarse sand to make coarse sandstone, or fine sand to make the fine-grained sandstones.

During this period, we know these valleys were deepening; for the beds are from three to six thousand feet thick, slowly sinking and slowly filling up with sediment as age after age passed away.

The fossils of the new red sandstone are not abundant. One reason may be, that sand is a very unfavorable material for the preservation of fossils; for in the Valley of the Connecticut we have evidence that life during this period abounded in numerous forms, though little more than the prints of their feet are left to tell the story of their lives.

Over a space in the Connecticut Valley ninety miles long, and from two to three broad, footprints of various animals have been found on slabs of sandstone, — evidently impressed when this was soft mud by the shore. They were first discovered by Dr. Deane, and carefully studied by Professor Hitchcock, who thought that the tracks then discovered indicated one hundred and nineteen species of animals, of which he regarded thirty-one as birds, seventeen lizards, eleven frogs, eight tortoises, four fishes, twenty-six articulates, ten bird-lizards, and five marsupialoids, or animals resembling those mammals that have a pouch for their young, such as the opossum.

The largest of the bird-like tracks is eighteen inches long, and the length of stride from two and a half to five feet. Professor Hitchcock calculated the height of the animal at twelve feet, and its weight from four to eight hundred pounds. From the depth of the impression made, the animal must have been very heavy. I have seen one track which holds a gallon of water.

On some slabs at Turner's Falls, the tracks are so abundant, that no one can be distinguished from another. They look as we may see a muddy road after a flock of sheep has passed.

The reptile-tracks are from one-fourth of an inch to twenty-two inches in length. The largest reptile seems to have been a kind of frog. A frog twenty-two inches long would be a monster; but what kind of a frog was that whose foot was twenty-two inches long, and from thirteen to fifteen inches wide? Professor Hitchcock thought it was "almost as heavy as an elephant." It was web-footed, and the front feet were not more than one-third as large as the hind ones. In Amherst cabinet is a slab thirty feet long, on which are eleven of these tracks.

Another reptile was a lizard, with a foot fifteen inches long, having a heel nearly as large as a horse's hoof, and armed with a stout spur.

Some of the impressions of insects are so fine, that

they require a magnifying glass to bring them out distinctly.

Certain of these impressions, which were regarded as undoubted bird-tracks, prove to have been made by some remarkable quadruped. I called upon Professor Hitchcock during his last illness, and found him in bed, propped up with pillows; while upon the floor before him was a slab containing a number of the so-called bird-tracks. The tracks of what every one might regard as a bird with three toes, the foot three or four inches long, were plainly visible on one portion of the slab. But, on following the tracks along, the animal had rested; and there was the impression of its long heels, making its track sixteen inches long; while in front of them were marks made by small fore-feet that had just touched the ground, while behind was a spot on the

Fig. 18.

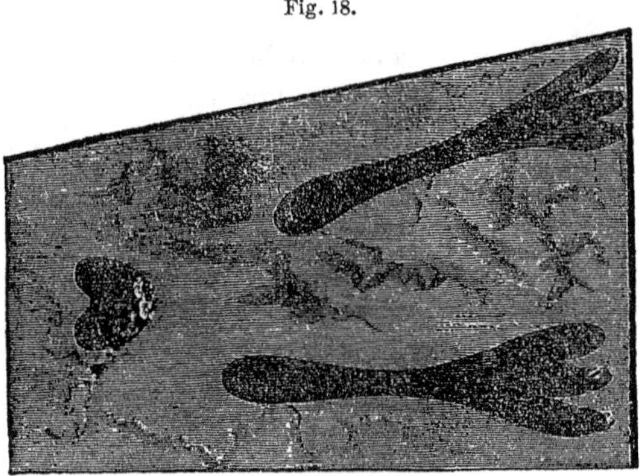

Anomœpus major.

slab where the tail had rested. Fig. 18 represents a portion of this slab, which is now in the ichnological cabinet at Amherst. If a bird, then, it had four feet, and a tail like a lizard. And what kind of a bird could that be? Professor Hitchcock said, that, in his opinion, some of the tracks were made by four-footed birds having some resemblance to mammals; while others were made by batrachian, or frog-birds. He says in one published statement, "I must believe that these animals combined characters now found distributed among birds, lizards, batrachians, and perhaps mammalia."

Only by slow degrees did life mount and fly. Prone on the water at first it lay, or crawled upon the ground; then, elevated slightly above it, the first reptiles crept. The fore-feet then were elevated, and the hind-feet alone were employed for progression. The fore-feet became small by disuse; and it is not unreasonable to suppose that from these, eventually, membranous wings were developed, and feathers completed the bird. True birds, probably, existed during this period, if not previously.

Strange glimpses of the by-gone world and its remarkable inhabitants these footsteps of the new red sandstone give us. Where the beautiful Valley of the Connecticut is, the cultivated field, and the crowded city, an arm of the ocean was, with its low, broad, flat, muddy shore, at times overflowed by the waves.

It is early morning; and the fog slowly rises from the land, unveiling a picture that dazzles us with its unrivalled beauty. Over the eastern hills appears the red sun, larger and fierier than to-day; the pine-like trees upon the mountain-tops waving their lordly crests. Stretching away to a long bay or gulf is a slope overgrown with palm-like and luxuriant tropical trees, from which dangle in the air, and stretch from bough to bough, vines clad with flowers of brilliant dye.

On the shore of the bay is a low, flat beach, fringed on the land-side with broad-leafed reeds with tufted tassels. This beach is from two to three miles wide, composed of fine micaceous sand, at times covered by the high tide, and at times baking in the sun's hot ray.

But what of the inhabitants of this old world,—this world beyond a thousand floods? Here they come, crawling, hopping, stalking, down to the shore, some of them as tall as young giraffes, and with bodies as large as oxen. Down they move in constant procession across the sands to fish in the morning light. The water is alive with fish, frogs, and nondescript reptiles, swimming, crawling, diving, and filling the air with their din, —that din made more horrible as these gigantic waders fill their crops with the crawling brood.

In "The Principles of Geology," Lyell gives us the key to these impressions on the sandstones of the Connecticut Valley. "When examining, in 1842, the extensive mud-flats of Nova Scotia, which are exposed at low tide on the borders of the Bay of Fundy, I observed not only the footprints of birds which had recently passed over the mud, but also very distinct impressions of raindrops. The sediment with which the waters are charged is extremely fine, being derived from the destruction of cliffs of red sandstone and shale; and, as the tides rise fifty feet and upwards, large areas are laid dry for nearly a fortnight between the spring and neap tides. In this interval, the mud is baked in summer by a hot sun, so that it solidifies, and becomes traversed by cracks caused by shrinkage. Portions of the hardened mud between these cracks may then be taken up, and removed without injury."

I have taken up many of these near Windsor, in the

Bay of Fundy, containing impressions of dogs and men that had walked over the spot, and prints of raindrops identical with those found on new red slabs, so common in many localities. It is a marvellous fact, that while millions of men have lived, and striven their best to leave an enduring record behind them, but have been swallowed in Oblivion's never-to-be-satisfied maw, the rain of millions of years ago, falling all unconsciously upon the sinking sand, — the very types of all that is perishable, — has left a record so enduring, that we read it now, and can tell by the shape of those rain-pits the direction of the wind that blew so long, long ago. I have seen a brick from Nineveh, more than two thousand years old, with the impression of the foot of a dog upon it, made, evidently, while it was yet soft as it lay in the sun to dry. I have a brick in my possession taken from the chimney of an old house, which bears a very distinct impression of a hen's foot. Such impressions might, of course, be preserved for millions of years, as those of the Connecticut Valley have been.

Near Richmond, Va., there are several seams of good coal in this formation. There are also others in North Carolina; and they show that coal-making did not end with the carboniferous period.

Fig. 19.

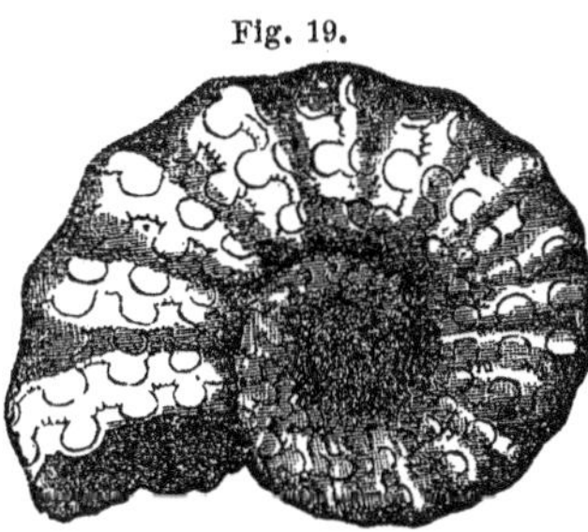

Ceratites nodosus.

Many of the plants were allied to those of the coal-measures. Calamites, equiseta, and ferns are the most common. Among shells, spirifers and *producti* are abundant; and, among cephalopods, the ammonite family begins to rise into importance. Fig. 19 represents one genus of this family, from the muschelkalk of Luneville, France. Fossil fishes are quite abundant in some localities, and most of them have heterocercal tails. Sunderland in Massachusetts is a famous locality for obtaining fossil fish of this period. The shales in which they are found contain considerable petroleum; and this petroleum may have been the cause of their death, and the means of their preservation as fossils.

Fossil reptiles have been found in this formation in North Carolina, Prince Edward's Island, and Phœnixville, Penn. In the *Bunter Sandstein,* or colored sandstone of Germany, and in Lancashire and Cheshire in England, many footprints of an enormous, frog-like reptile have been found, called originally *cheirotherium,* or hand-beast; for the impression made by the foot of the animal resembled that of a rude, human hand. One slab containing them (Fig. 20), from the new red sandstone at

Fig. 20.

Cheirotherium Barthi.

Jena in Germany, is now in the Ward Museum, Rochester, N.Y. The tracks of the hind-feet are about eight inches long and five wide; and the fore-feet, four inches long and three wide. Remains of this animal have since been discovered; and it is now known as the *labyrinthodon*, from the labyrinthine appearance of the interior of the teeth. Fig. 21 represents the head of one found near Stuttgart, Wurtemberg. The body has not yet been discovered, but, from the size of the head, is estimated at nine feet in length. It was covered with scales. This reptile first made its appearance in the carboniferous formation, but was then smaller.

Fig. 21.

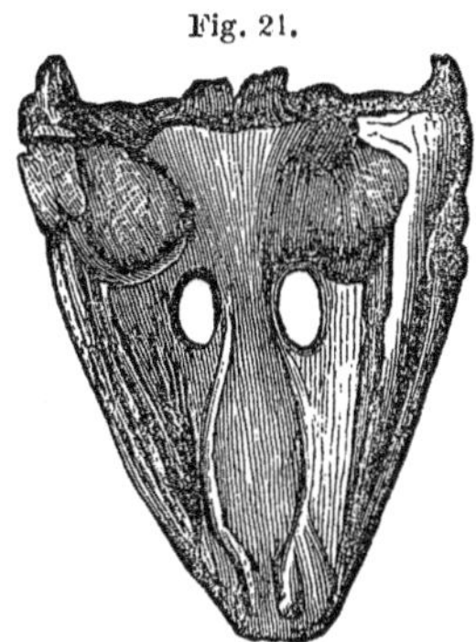

Labyrinthodon Jægeri.

Another remarkable reptile of this time, of which but the head has been discovered, is the *placodus*, or plate-tooth, as its name signifies. Its crushing teeth were like paving-stones. It was discovered in the muschelkalk, at Laineck, in Bavaria, associated with multitudes of fossil shells, which have given their name to the strata; and there is little doubt that the teeth were used, being well adapted for that purpose, to crush the shells of the mollusks on which the reptile fed. Of the last tooth, says Owen, " It is, in proportion to the entire skull, the largest grinding tooth in the animal kingdom."

In triassic beds of England, Germany, and in coal-beds in North Carolina, supposed to be of this age, we find the first mammals that have yet been discovered. They are all small in size. From the shape of their teeth, they appear to have been insectivorous, that is, insect-eating; and, according to Owen, they were probably marsupial, that is, had a pouch for their young, like the opossum. This fact is one well worthy of remembrance; these earliest known mammals, from such widely-separated localities, being all referred to a type, the marsupial, which is that type of mammals most closely allied to birds in its organization, and belonging to the lowest order of the class.

The *Bunter Sandstein* is a fine-grained, whitish sandstone, some of it nearly as coarse as conglomerate, and used for making millstones. Occasionally, beds of marl are found associated with it. It is spread over a large portion of France and Germany, and in many districts contains numerous remains of fossil plants and marine shells.

The *muschelkalk* is a compact limestone of a grayish color, and abounds with the remains of radiates, shells, and fishes. The ocean in which it was deposited appears to have been very muddy, and therefore unfavorable to the growth of coral. The polyps could find no solid foundation on which to build; but, for that reason, it was eminently favorable for the growth of crinoids, in which it abounds. One species, termed the lily encrinite, is composed of twenty-six thousand separate pieces.

The remains of fishes are most abundant, consisting mostly of teeth and scales: the teeth, round and flattened, were distributed over all parts of the palate, and were just suited to the work of crushing shells, such as abound in this formation, and from which it receives its name.

The triassic period I regard as one of the most remarkable in the world's eventful history. Fishes abound; reptiles are more common and larger than in any previous time; birds are numerous and gigantic; and mammals appear. We may expect much light to be thrown upon the early history of birds by a more careful study of the footprints of the Connecticut Valley; and shall probably discover many linking forms between birds and lizards, which seem to have existed in the greatest abundance during this time. I think it, however, highly probable that the beds of the Connecticut Valley will be found to be much more recent than has been generally supposed.

LECTURE IV.

By this time, there are some explanations necessary to clear the way that lies before us, and to make more distinct and accurate our knowledge of that already gone over. I have spoken of the granite as the oldest rock of all; but this, although true in the main, is not absolutely so. As the interior of the earth is cooling to-day, so granite down there is forming to-day, as the rock formed from that cooling, slowly, crystallizes. This is equally true of all the geologic periods; and we have therefore granites of all ages, and frequently find veins of it ramifying through the more recent formations.

I have also said that the metamorphic rocks were the next formed, and that they were laid down at a time when no life existed on the globe. This statement, though generally correct, requires to be modified. As metamorphic rocks are those sedimentary rocks which have been altered by heat, we have metamorphic rocks of all ages, because, in all ages, rocks have been exposed to such heat as would, when cool, cause their particles to be re-arranged and crystallized. In Canada, the rocks belonging to the age of radiates, the oldest rocks in which the remains of living beings are found, are so changed by metamorphic action, that it was years before the fossils abounding in some of them were discovered. In New England, we have in many places metamorphosed Silurian and Devonian rocks. The limestones, being melted, crystallized on cooling, and produced marble, in which fossil forms are completely obliterated; while the original shales have hardened into slates. In Eastern Pennsylvania are metamorphosed coal-measures, the coal in which has been transformed by heat from bituminous to anthracite; which is, therefore, metamorphic coal. In other parts of the world, we find metamorphosed strata up to the more recent tertiaries. Some of the beds that we thus find are only partially metamorphosed, and fossils can be distinguished in them; while in others, thoroughly done, all of life that the eye can trace is gone.

From what has been said, some might suppose that all the rocks, from the granite up, lay in regular order one upon another, like the coats of an onion; but this is far from being so. Let the rocks be represented by the numbers one, two, three, four, &c. We frequently find, in fact invariably find, some of these numbers wanting. Number three is found lying on number one; number two being absent. Number nine is found on number four; all between being absent. In other cases, formations which are ten thousand feet thick in one country are reduced to a thickness of forty or fifty feet in another. Thus, in Portugal, the coal-measures rest upon the granite; and the Metamorphic, Silurian, and Devonian are wanting. On the shores of Lake Erie, glacial beds rest upon Devonian rocks; and all the formations lying between the Devonian and the drift are wanting. In Middle Tennessee, the Devonian formation, which in New York and Pennsylvania is several thousand feet in thickness, is reduced to a thickness of from thirty to fifty feet; while all the Silurian rocks above the Trenton limestone are wanting.

There are three reasons that may be given for this condition of things. One is, that those parts of the earth where particular beds are wanting were dry land at the time when such beds were laid down at the bottoms of lakes or oceans: hence no sediment was deposited there, and no beds were formed; instead of that, rains swept over them, and carried off sediment, to make deposits in other localities covered by water. Thus we may suppose that those portions of Portugal where the coal-measures rest on the granite were mountain-lands at the time that the Silurian and Devonian beds of Great Britain and the United States were forming at the sea-bottom. After the deposition of the Trenton limestone, Middle Tennessee was probably elevated above the water-level; while in New York and part of Canada, the Hudson-river group, the Niagara limestone, and the other upper Silurian beds, which are wanting in Tennessee, were being deposited.

Or those portions of the earth may have been covered with the waters of a deep sea, far from land and sediment-bearing rivers: for, in the depths of the Atlantic and Pacific, there cannot be much sedimentary deposition or deposits of animal forms, save those of minute infusoria; and in future time, should it become dry land, the beds that represent this age will be found principally where the ocean is shallow; where shells, fishes, and corals abound; and near to rivers, which are constantly sweeping in sediment.

Or, lastly, beds may have existed representing those periods, and been swept off by denudation since that time. Thus, in Scotland, Lyell tells us of a great body of sandstone, from one to three thousand feet in thickness, covering a large part of Ross-shire, that has been removed, — swept off by rains, rivers, or currents, or all combined. "Professor Ramsay," he says, "has pointed out considerable areas in South Wales and some of the adjacent counties of England, where a series of primary strata, not less than eleven thousand feet in thickness, have been swept off." When we think that some rocks must have been exposed to the action of the elements for millions of years, the amount of denudation must be great: in fact, the thickness of any formation is an accurate measure of the denudation of the previously existing rocks; for the world has been built up as the modern Egyptians build their cities, — the ruins of the old supplying material for the new.

Lias. — The name "lias" is an English one, and was probably given to the group of rocks lying above the trias, from the fact that it consists of a number of thin layers of limestone, marl, and shales; "lias" being, probably, a corruption of layers. Its thickness is from five hundred to a thousand feet, and it abounds throughout with many remarkable fossils.

Shakspeare has divided the life of man into seven ages; and, not improperly, the organic period of the world's history may be divided into a like number. The age of radiates: Before the first shell clothed the first ocean-tenant, in the heated waters floated the radiates, crawled on the sea-bottom, expanded their flower-like cups, and, without intermission, silently built their stony habitations in the shallow seas. The age of shells or mollusks: Then shells paved the bed of the sea, and accumulated till islands of them rose from its depths. The age of fishes: When bony-plated fishes by myriads lashed the waters of all seas, and their accumulated remains helped to build the foundations of continents. The age of plants: When, wherever land was, or marsh, plants flourished abundantly, and green forests waved over equator and poles alike. The age of reptiles: When reptiles crawled on the land, swam in the water, dived into its depths, and, on leathery wings upborne, tenanted even the very air. The age of mammals next: When whales and seals crowded each other in the waters; and beasts, from the mouse to the mastodon, roamed over the land. Last, the age of man, in whose spring-time we live.

Fig. 22.

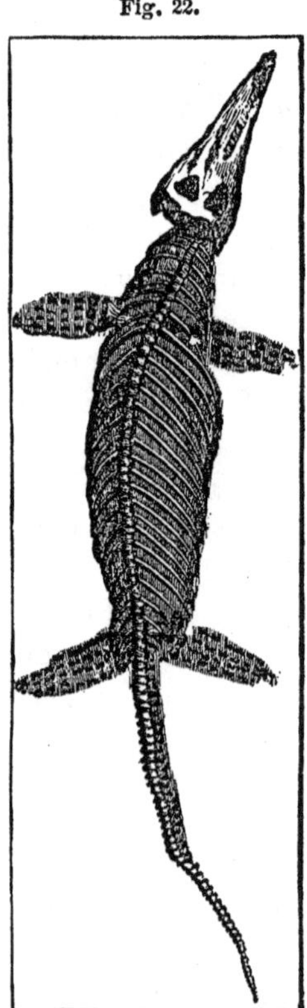

Ichthyosaurus intermedius.

During the liassic period, the ocean especially seems to have swarmed with reptilian forms. One of the most remarkable was the *ichthyosaurus*, or fish-lizard, as its name means (from the Greek *ichthys*, "a fish," and *sauros*, "a lizard"). Fig. 22 represents a fine specimen from the lias near Glastonbury in England, and now in the British Museum. It was as large as a young whale; some of the largest measuring from thirty to forty feet in length. We find in it the snout of a porpoise combined with the teeth of a crocodile, the head of a lizard with the backbone of a fish and the breastbone of an ornithorhynchus with the paddles of a whale. The moutn was wide, the jaws long, the teeth conical, and much like those of a crocodile, but more numerous. In some species, a hundred and eighty have been counted. As they were liable to lose them, new ones were pushed up from below to take their place.

The eye was of enormous size, sometimes larger than a man's head. Fig. 23 represents a head of the ichthyosaurus from Lyme Regis, England, in which the eye is seven and a half inches in diameter: it was protected by a series of thin, bony plates. From its great size, this animal must have been able to descry his prey in the twilight depths of the ocean or in the obscurity of night; for these monsters lived by preying, as their remains abundantly testify. Within their ribs, where the stomach was placed, have been found masses of half-digested bones and scales of fishes that lived in the same seas and at the same time as the ichthyosaurus. We can tell not only the food of the animal, but the size and shape of the stomach and intestines. The stomach formed a pouch, or sac, extending through nearly the entire cavity of the body. Bones of small ichthyosauri have been found within the bodies of large ones, the individuals to which they belonged severa' feet in length.

Fig. 23.

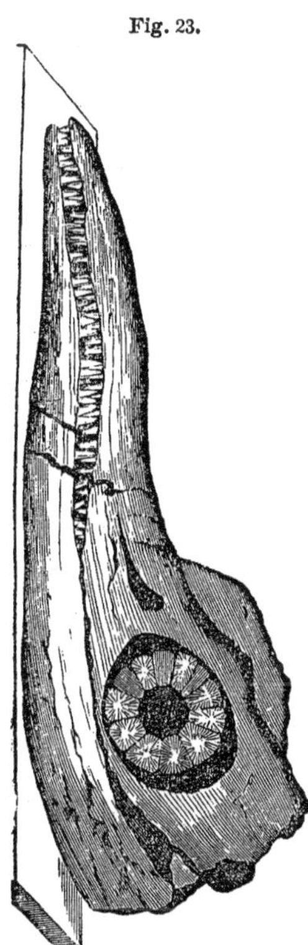

Ichthyosaurus platyodon.

It is a singular fact that their excrement is found in great abundance; and, under the name of "coprolites," they are gathered, ground up, and used for fertilizing English soils. In them are also found in abundance the teeth and bones of fishes which have passed through the bodies of these reptiles undigested.

Such fossils as these are of the greatest value to the geologist. They enable him to fortify his science so as to make it impregnable against the attacks of those who teach that the world was made "just as it is." I saw a Cornishman digging for lead in Wisconsin, and asked him where he supposed the shells came from that abound in the limestone there.

"Why," said he, "the Lord made them when he made the rocks." Some men who are pretenders to science have lent their names to the support of a similar idea. They argue thus: "There is no reason to believe, as geologists teach, that these buried trees, shells, fishes, and beasts ever had an actual existence. Could not the God who made the myriad trees that adorn the earth to-day make the appearance of trees in the rocks when the world was made? Could not He who made the shells that spangle our seacoasts, and the fishes that swim in the ocean-depths, supply the rocks with the rude appearances of these at the beginning? All things were made by the Omnipotent, when out of chaos the creative fiat brought a perfected world."

Five minutes' examination of a fossil ichthyosaurus should be sufficient to show the absurdity of this conclusion. Look at these teeth, long and sharp, for holding the slippery prey! Some, as you see, are worn by use, and others broken; while in this young one they are all sharp and perfect. Within its ribs is a bony mass. We break it, and there are the glittering scales of the fishes it fed upon, — the remains of its last breakfast; while the conrol tes below bear the impression of the internal surface of the intestines in which they once lay, and in them we may see the undigested scales and teeth of the fishes and reptiles that it had swallowed. Are we to be told that this was created "just as it is" when the world was made? The man that can believe that must have a larger swallow than the ichthyosaurus itself, and is quite as much of a fossil. He should be ticketed, and laid away on the shelf of some museum for the consideration of the curious in the year 2000, when the existence of such beings may be denied.

"But is it not possible that these were created just so?" I answer, No. It is not possible in the nature of things. There is nothing like it in Nature: she does not deal in shams and make-believes. What lies in stone these fossils would be! and for no purpose but to puzzle and perplex poor mortals. And where could we draw a line between the real and the unreal?" Are the Pyramids actual structures reared by man? or were they, too, made when the world was made? Here is an Indian mound. On opening it, we find arrow-heads and spears of flint, a hatchet of greenstone, a pipe, and one of those sugar-loaf-shaped skulls that indicate the ancient mound-builders. We commence to solve the problem lying before us. Here is the skull of what was once a man; for there is a marked distinction between the shape of the male and female head. He was evidently no great reasoner; for the reflective portion of the head is low, and the skull over that region thick. He had some constructive talent, as these arrow-heads and other implements indicate. He did not know the use of iron, evident ly: he did know the use of tobacco. Poor savage! As we are thus reasoning, up comes our world-made-as-it-is friend. "Nonsense!" he exclaims. "The God that made the mighty mountains — could he not make a little mound like this? He who made man at the beginning could surely make the appearance of a skull. He who created the millions of grassy spears at our feet, and the tobacco of the field, might surely make these spears, hatchets, and this semblance of a pipe. They are merely types, made at the creation, of what was destined to appear in due time."

It is just as reasonable to suppose that God made tobacco-pipes that men never saw or smoked as that he should have made fishes and trees in the rocks that never lived or grew. Admit such reasoning, or rather the want of it, as such talk indicates, and we should be bounded by the limitation of our senses, and become the veriest infidels the world ever saw.

Thirty species of the ichthyosaurus are known. It was most abundant during the time of the lias and oölite.

Another marine reptile of the lias was the *plesiosaurus* (near to a lizard), from the Greek words *plesion*, "near to," and *sauros*, "lizard." It was so called because nearer to a lizard in its organization than the ichthyosaurus.

Fig. 24 represents a fine specimen from the lias at Glastonbury, England. *Dolichodeirus*, its specific name, means "long-necked." The vertebræ are not double concave, as in the ichthyosaurus and in fishes; nor is it concavo-convex, hollow on one side and rounded on the other, as in crocodiles, but nearly flat: and in this respect, as well as some others, it resembles the land-saurians. This reptile was from eight to twenty feet long. It has the head of a lizard, with more than a hundred teeth like those of a crocodile; a neck resembling the body of a serpent, having thirty-three vertebræ; while living reptiles only have from three to eight. Its trunk and tail resemble those of a quadruped. The tail was short, but the paddles long.

Fig. 24.

Plesiosaurus dolichodeirus.

In the fore paddle of the plesiosaurus are all the essential parts of a human arm, — the scapula, humerus, radius, and ulna, the bones of the carpus and metacarpus, and the phalanges for five fingers, — a prophecy (was it not?) of the wonder-working arm and hand of man.

With a long, arched neck, like that of a swan, in the sheltered bays of those old oceans gracefully paddled the plesiosaurus, darting down its small head, and catching the finny prey that constituted its food; for within its ribs have been found the remains of the fishes on which it fed.

The most extraordinary animal of this time, and I had almost said of any time, was the *pterodactyle* ("wing-finger"), from the Greek words *pteron*, "a wing," and *daktulos*, "a finger."

Fig. 25.

Pterodactylus crassirostris.

Fig. 25 represents the most perfect skeleton of the pterodactyle ever found. It was discovered in the lithographic limestone at Solenhofen, Bavaria, associated with the remains of dragon-flies. Its specific name, *crassirostris*, means "thick-beaked."

Cuvier, in his "Osteology," gives the following description of this bird-like reptile: "You see before you an animal, which, in all points of bony structure, from the teeth to the extremity of the nails, presents the well-known saurian characteristics, and of which one cannot doubt that its integuments and soft parts, its scaly armor, and its organs of circulation and reproduction, were likewise analogous. But it was at the same time an animal provided with the means of flying; and, when stationary, its wings were probably folded back like those of a bird; although, perhaps, by the claws attached to its fingers, it might suspend itself from the branches of trees. Its usual position, when not in motion, would be on its hind-feet, resting like a bird, and with its neck set up, and curved backwards, to prevent the weight of the enormous head from destroying its equilibrium. The animal was, undoubtedly, of the most extraordinary kind, and would appear, if living, the strangest of all creatures."

Cuvier was mistaken in supposing that it had scaly armor: it was more bird-like, and hence stranger than even he supposed.

It has four fingers, terminated with long hooked claws, of about the length that an animal of its size might be expected to have; but its fifth finger, occupying the place of a little finger, is inordinately large and long. From this long finger a web extended down the side of the animal's body, by means of which it could fly. The form of its neck and the hollowness of its bones resemble those of birds; its wings approach those of the bat; its body and tail are like those of a mammal; while its beak is furnished with sixty sharp, conical teeth. Thus one naturalist considered it to be a bird; another a kind of bat; and a third regarded it as a flying reptile, which it is now generally supposed to have been, yet having a very bird-like organization.

It strongly reminds one of Milton's description of Satan making his way from Pandemonium to Paradise :—

> "The Fiend,
> O'er bog or steep, through strait, rough, dense, or rare,
> With head, hands, wings, or feet, pursues his way,
> And swims, or sinks, or wades, or creeps, or flies."

So the pterodactyle waded, crept, walked, swam, dived, and flew, equally at home in water, in air, or on land.

Being found associated with insects, it has generally been regarded as an insectivorous reptile ; but those strong jaws, furnished with long, strong, conical teeth, are no mere fly-nippers. An animal furnished with such powerful claws and teeth must have been a rapacious monster, from which but few animals could escape. Diving in the water, it seized the darting fish ; pounced upon struggling reptiles on shore ; the labyrinthodons with their scaly bodies no doubt falling at times victims to its terrible attacks : nor could birds fly more rapidly than those powerful wings could follow.

Professor Owen describes three large species of pterodactyle, measuring fifteen feet from the tip of one wing to the tip of the other. He also says that another species, *Pterodactyle Cuvieri,* was probably upborne on an expanse of wing of not less than eighteen feet. Recently some fragments of pterodactyle bones have been discovered near Cambridge, in England, which Dr. Buckland, jun., considers must have belonged to an individual that measured twenty-seven feet. Thirty-seven species of pterodactyles have been identified and described.

Stranger monsters than fable ever fancied peopled the young world ; and stranger, perhaps, than any yet discovered remain to be found : we have but read a few torn leaves of a large volume.

Fossil vegetable remains are common, — such as ferns, cycads, and conifers, or cone-bearing plants. The cycads are plants resembling palms in their outward appearance, and cone-bearing trees in their inward structure. From the abundance of their leaves and trunks found in a fossil state in England, they must have formed at one time a large part of the vegetation of Great Britain.

Insects are very common in some of the limestones of this period: several hundred specimens have been discovered in liassic beds, which bear a strong resemblance to living insects. One band of limestone in this formation, found in Gloucestershire, England, has been called insect limestone, from the abundance of their remains found in it.

Mollusks were numerous, and some of them of great interest. Figs. 26 and 27 represent two common forms

Fig. 26.

Ostrea Marshii.

Fig. 27.

Gryphea arcuata.

found in the liassic beds of Europe. The tropical seas of this time abounded with cephalopods, the highest class of mollusks. Among them is the *nautilus*, the *ammonite*, and species resembling the cuttle-fish. The modern cuttle-fish is provided with an ink-bag; and, when pursued by an enemy, it darkens the water with the ink, and, under the cloud thus formed, escapes. Fossil ink-bags similar to these have been found in the lias, showing that this is a very ancient contrivance. The animal has been drawn by the fossil ink thus furnished, which is said to equal India ink. Notwithstanding this contrivance, coprolites of the ichthyosaurus have been found stained with their ink, showing that they were preyed upon by these merciless marauders.

Fig. 28.

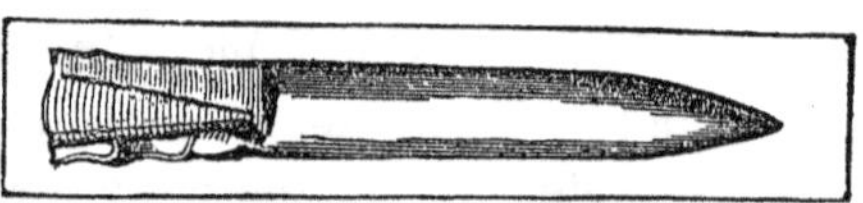

Belemnite Owenii.

The *belemnite* (stone-dart), as one of these cuttle-fish is called, from the dart-like shape of its internal bone (Fig. 28), is very abundant in the liassic and cretaceous beds. More than a hundred species have been described.

Fig. 29.

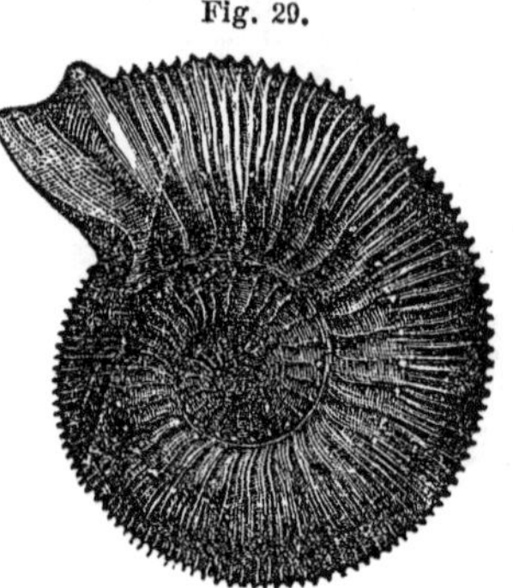

Ammonite Humphresianus.

The *ammonite* (Fig. 29) derives its name from its resemblance to the horn of Jupiter Ammon. They are found from the size of a cent to the size of a cart-wheel, and are frequently called snake-stones. Scott refers to them in his "Marmion:"—

"Thus Whitby's nuns, exulting, told
How that, of thousand snakes, each one
Was changed into a coil of stone
When holy Hilda prayed:
Themselves within their sacred bound
Their stony folds had often found."

Being very abundant in the liassic beds at Whitby, in Yorkshire, England, they are collected and sold to visitors, who sometimes ask where their heads are. The collectors, ready to oblige their customers, file one end of the shell into the shape of a head; and so strong is the resemblance, that he must be an unreasonable sceptic who denies that they are Lady Hilda's veritable converted snakes.

The shell of the ammonite is thin, and could not therefore endure much pressure; but it is strengthened by ribs and tubercles, which add to its beauty no less than its strength. The septa, or divisions of the shell, ramify into each other at the point of junction. "Nothing can be more beautiful," observes Dr. Buckland, "than the sinuous winding of these sutures in many species at their union with the exterior shell, adorning it with a succession of most graceful forms, resembling foliage or elegant embroidery; and, when these thin septa are converted (as they sometimes are) into iron pyrites, the edges appear like golden filigree-work, meandering amid the pellucid spar that fills the chambers of the shell."

Fig. 30.

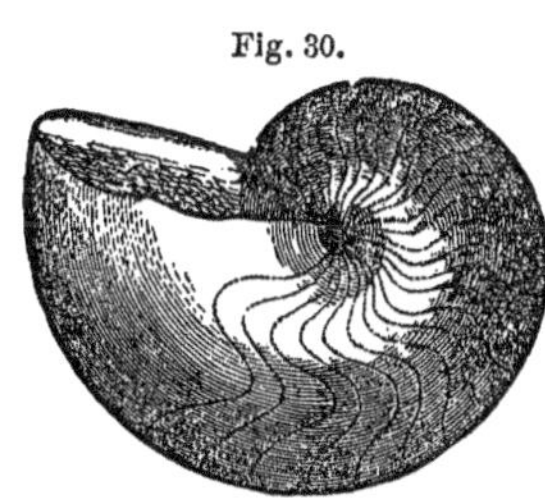

Nautilus pseudo-elegans.

The nautilus (Fig. 30) (from the Greek *naus*, "a ship") was an allied form inhabiting the same oceans. Its shell is generally rounder and the whorls less numerous than in the ammonite. After the ammonite perished, the nautilus still continued to flourish; and, in our present tropical seas, two or three species still exist. G. F. Richardson of England has embodied this fact in beautiful verse, which I have taken the liberty to change slightly: —

"The nautilus and the ammonite
Were launched in storm and strife;
Each sent to float, in its tiny boat,
On the wide, wild sea of life.

And each could swim on the ocean's brim,
And anon its sails could furl,
And sink to sleep in the great sea deep,
In a palace all of pearl.

And theirs was a bliss more fair than this
That we feel in our colder time;
For they were rife in a tropic life,
In a brighter, happier clime.

They swam 'mid isles whose summer smiles
No wintry winds annoy;
Whose groves were palm, whose air was balm,
Where life was only joy.

They roamed all day through creek and bay,
And traversed the ocean deep;
And at night they sank on a coral bank,
In its fairy bowers to sleep.

And the monsters vast of ages past
They beheld in their ocean caves:
They saw them ride in their power and pride,
And sink in their billowy graves.

Thus hand in hand, from strand to strand,
They sailed in mirth and glee, —
Those fairy shells, with their crystal cells,
Twin-daughters of the sea.

But they came at last to a sea long past;
And, as they reached its shore,
On the storm-wind's breath came the blast of death,
And the ammonite lived no more.

And the nautilus now, in its shelly prow.
  As o'er the deep it strays,
Still seems to seek in bay and creek
  Its companion of other days.

And thus do we, on Life's stormy sea,
  As we roam from shore to shore,
While tempest-tost, seek the loved, the lost,
  But find them on earth no more.

Yet the hope how sweet! — again to meet,
  As we look to a distant strand,
Where heart meets heart, and no more they part
  Who meet in that better land."

There is poetry in geology; and I am glad that some one has eyes to see it, and ability to delineate it. Geology is not that dry, dusty science that some suppose, but instinct with life and beauty in every part; and what are thus naturally joined together neither lecturer nor writer should put asunder.

Oölite. — The oölite derives its name from small egg-like bodies, resembling the roe of a fish, that are found in some of its limestones (*oön* is the Greek for "egg"). On breaking these small round bodies, we find in the inside of them a fragment of shell or a grain of sand; and, around this, limy matter has gathered into a small ball. These appear to have been made very much as a confectioner makes comfits. The seeds that he intends to coat he places in a pan, which is constantly moved backward and forward, so as to keep the seeds continually rolling while melted sugar is dropping upon them. In this way, they become pretty evenly coated with the sugar; and the comfits are made, each enclosing a seed. Nature made comfits on a much grander scale: the ocean was the pan, fragments of shells and the grains of sand were the seeds, the waves of the constantly-rolling sea supplied the motion, while the lime contained in the water coated them with concentric layers.

In England, these rocks cross the country from Yorkshire in the north-east to Dorsetshire in the south-west, in a belt having an average width of about thirty miles. Along the Yorkshire coast, they form a large part of the sea-wall; and the fossils contained in them are picked up in great abundance.

The beds of the lias and the oölite are found on the Jura Mountains in the west of Switzerland, and together are frequently termed Jurassic beds; and both formations are then included in the Jurassic formation, which, on the Jura Mountains, consists of beds of whitish or yellowish limestone.

At Solenhofen, in Germany, there are quarries in the oölitic formation which produce a fine-grained limestone, which is used for lithographic purposes. A drawing being made upon the smoothed surface of the stone with a peculiar kind of ink, impressions are taken from it by a printing-press, just as from a wood-cut or from type. From these quarries have been obtained an immense number of fossil fishes, crustaceans resembling shrimps and lobsters, echini or sea-eggs, insects, reptiles, and reptile birds. It has been supposed that this limestone was deposited on a coast where the water was shallow, and on the shores of which flying reptiles chased their prey.

Among the animals found in the lithographic stone at Solenhofen are many species of the pterodactyle and several saurians; but the most remarkable fossil is one which has been recently discovered, and termed the *archeopteryx.* It was probably a feathered reptile; but the head of the animal has not yet been discovered. Its toes are all armed with sharp claws. Its tail, six inches long, consists of twenty vertebræ, with a pair of feathers attached to each vertebra. It appears to be one of those strange linking forms between one class and another such as some of the Connecticut-valley footprints indicate, many of which probably are yet to be found.

At Stonesfield, in England, have been found several jaws and teeth of small mammals. They are considered marsupial by Owen, and some of them insectivorous. They most resemble some of the marsupial forms of Australia.

Owen refers to the interesting correspondence between the organic remains of the oölite and the existing forms of the Australian continent and the surrounding seas. The *cestracion*, or Port-Jackson shark, with its palatal, crushing teeth, has given us the key to the larger cartilaginous fishes of the oölite, which had similar teeth. A recent species of *trigonia* has been discovered on the Australian coast; and the trigonia (Fig. 31) is found in the lower oölite, Wurtemberg. Even the vegetation of Australia bears some relation to that of the oölite, as its araucariæ and cycadeous plants testify.

Fig. 31.

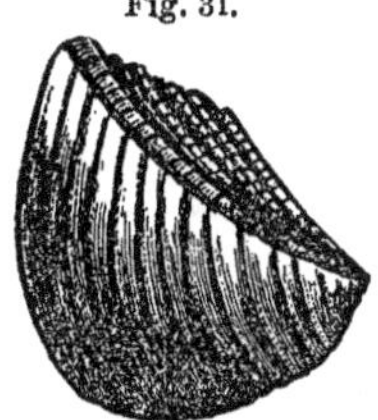

Trigonia costata.

There is a remarkable bed in the Isle of Portland, England, known as the Portland "Dirt Bed," which is the soil of that ancient period, buried and hardened, but still retaining so much of its primitive condition as to be readily recognized. It is about a foot thick, and is made up of black loam mixed with the remains of tropical plants, sticks, seeds, leaves, mingled and compressed together. Partly sunk in this black soil, and partly covered by overlying slate, are silicified trunks of large coniferous trees, some of them more than thirty feet long, and stumps also silicified, at nearly the same intervals at which trees are found growing in a modern forest, with their roots attached to the earth in which they grew. They are from one to three feet long and are mostly erect.

The remains of more than a dozen species of small mammals have been found in this "dirt bed." They are generally regarded as marsupial.

Wealden. — In the south-eastern part of England, there is a series of deposits, principally of fresh water, sometimes classed with the oölite, consisting of limestones, marls, sands, clay, and shales. Occurring as they do in the wealds, or woods, of Sussex and Kent, they are known as the Wealden Group, and abound with remains of plants, insects, fishes, and peculiar reptiles.

The most remarkable fossils of the Wealden are fragments of reptiles which have been discovered distributed through beds in Sussex, Kent, and in the Isle of Wight. Dr. Mantell, in examining various quarries through these parts of England, discovered fragments of bone which could not be attributed to the skeleton of any known animal. They belonged, as he supposed, to a reptile; for the bone-cells in reptiles' bones are different in shape from those in birds and mammals, so that the smallest fragment can be distinguished. But the enormous size of the animal indicated by these bones seemed to be too great for belief; for who could credit the existence of a reptile whose body was larger than that of an elephant? At length, one fragment after another having been discovered, it could be no longer doubted that an enormous herbivorous lizard had once existed in

England, — so large, that a perfect thigh-bone has been found three feet eight inches in length; while some fragments indicate that this bone attained a length in some of nearly five feet, and, with the muscles and integuments, must have formed a limb eight or nine feet in circumference. If the leg-bones were of equal length, what a colossal reptile was this! — the top of whose back a man on horseback could not have reached. Its teeth resemble those of a lizard found in the West Indies, and called iguana: hence this animal is called *iguanadon* (pronounced igwaunadon), or iguana tooth. The teeth indicate a vegetable feeder: hence it must have had a bulky body. A model of one placed in the Crystal Palace at Sydenham, and there is good reason to believe not a whit larger than some of the originals, held twenty-one gentlemen, who took dinner inside; Dr. Owen sitting in its head for brains.

A strange time was this reptile age. The battle that had been waged in the ocean for ages was now extended to the land. Enormous lizards wandered through tropical woods, feeding upon plants and soft-bodied trees: others, making up, in strength, agility, and ferocity, what they lacked in size, fed, in turn, upon them and on each other. Reptiles in the ocean, swimming and diving, their huge paddles leaving a wake behind them like a steamboat; others on the shore, crawling, or basking in the sun on the heated rocks. Some have naked skins; others glitter with scaly coats of impenetrable mail; while bony fringes ornament still others in serrated rows along the spine. Stranger still, here are reptiles with wings; some clad with feathers, that mount and riot in air. The whole world was given up to cold-blooded monsters; and for immense periods it must have seemed as if these were the highest existences destined to dwell on its surface.

The *megalosaurus* (large lizard) sometimes attained the length of thirty feet. This reptile was carnivorous, furnished with teeth admirably adapted for their office. Dr. Buckland says of them, "In the structure of these teeth, we find a combination of mechanical contrivances analogous to those which are adopted in the construction of the knife, the sabre, and the saw." Each tooth had a double serrated edge of enamel; so that it cut "like the two-edged point of a sabre, equally on each side." The older teeth are curved backward like a pruning-knife; and the convex portion is thick, as the back of a knife is made thick to increase its strength. These teeth are sometimes found broken, no doubt in the terrible conflicts that took place between it and its scaly prey; and the young tooth is frequently found in the jaw, ready to take the place of the old one when its work is done. No perfect skeleton of this rapacious monster has yet been found; but, as Dr. Buckland says, so many perfect bones and teeth have been discovered, that we are nearly as well acquainted with it as we should have been if the whole skeleton had been found. The poet must have had such a reptile as this in his mind's eye when he wrote, —

"See! — late awaked, emerging from the woods,
He stretches forth his stature to the clouds,
Writhes in the sun aloft his scaly height,
And strikes the distant hills with transient light.
Far round are fatal damps of terror spread:
The mighty fear, nor blush to own their dread.
Large is his front; and, when his burnished eyes
Lift their broad lids, the morning seems to rise."

The *hylæosaurus*, or wood-lizard, as its name implies, was probably an herbivorous (herb-eating) reptile, from twenty to thirty feet in length. It was furnished with a crest along the back, composed of bones, some of which have been found seventeen inches long, and five inches broad at the base. Some existing lizards have small cartilaginous or gristly fringes of a similar kind; but these lizards of the olden time were as much superior to living ones in this respect as they were in size and power.

I am frequently asked, How is it, if the earth was advancing during all these geologic ages, that such inferior forms of life exist now, compared with those of the past? There are no such shells as the gigantic cephalopods of the Silurian seas; no such club-mosses or horse-tails, no such tall reeds as waved in the carboniferous swamps: our largest living reptiles are dwarfs compared with those of the oölitic times. How is this? The answer is, That the earth has been advancing as an abode of superior existences. The Silurian period was one better fitted for shell-life than any other; and hence, during this period, shells attained their greatest size, and were most abundant. The Devonian period was one better fitted for fish-life: hence fish multiplied, and shells dwindled. The reptile period was one in which the world was better fitted as an abode for reptiles than any other life-forms: hence their gigantic size and immense numbers during this period. But, when the earth became prepared for mammals, it was less favorable for reptiles (the conditions that best suited them having passed away), and beasts grew and multiplied; and, as it becomes best fitted for man's abode, the beasts dwindle, and man continually improves.

From the discoveries made in the Wealden beds, it

seems evident that a river like the Mississippi rolled over a continent, a large portion of which, probably, now lies beneath the bed of the Atlantic, sweeping into an estuary where the south-east of England now is, carrying down the spoil of the forest, and the remains of the various creatures that dwelt on its banks or floated in its waters. The sea has gathered, again and again, islands and continents into this their common grave; again and again to rise, and be crowned with life more triumphantly than before.

CRETACEOUS PERIOD.

Above the Wealden come the cretaceous beds. *Creta* is the Latin word for "chalk;" and this formation derives its name from the abundance of chalk found in it. The white chalk-cliffs of England have been noted from the earliest times. Like giants, they stand around the coast, guardians of the old land. They are portions of a great chalk-bed extending over a large part of Europe.

Whence came this soft, white rock, called chalk, with which we are all well acquainted? At first sight, it seems a difficult problem with which to deal. Geology, however, solves it. On close examination, we find chalk to be another of those wonderful productions formed in the great laboratory of life.

When writing on a blackboard, I have sometimes noticed a hard, offending particle that scratched the board; and, on carefully whittling with a pen-knife, have discovered a perfect shell that had been enveloped in the chalk. Large, perfect shells are frequently found, sometimes spiny shells; every spine perfectly preserved in the soft mud, now chalk, that surrounds it. But the fossils found in chalk in this way are few in number compared with those that are to be discovered in it. Take a tooth-brush, and brush into a teacup powder from a piece of chalk; wash it, pouring out the turbid water until you have the clean sediment at the bottom: on examining this, you may see with the naked eye small shells, spines of the echinus, spiculæ, or small spines of sponges, and fragments of corals. But what you can thus behold is but a small portion of what it really contains. A drop of the turbid water laid upon glass, and evaporated, leaves a white stain upon it. On this place a little Canada balsam, and heat over a spirit-lamp; and under the microscope you may then behold multitudes of shells, and fragments of shells, and other organic forms too small to be visible to the naked eye. Ehrenberg calculates that in one cubic inch of chalk there are a million and a half perfect fossil shells. Little think those ladies who powder their faces with prepared chalk, that, if we had microscopic eyes, we should see their fair cheeks looking like the beach of a tropical sea when the tide has gone down, — strewed all over with beautiful shells and corals. But perhaps it is best for all parties that we possess no such eyes.

At the bottoms of lagoons and basins in the West-India Islands, especially the Bermudas and Bahamas, a soft, white mud is found, which is produced by the wearing-down of corals by the waves washing over them, and the fecal matter of coral-eating fish. Darwin tells us of fishes seen through the clear waters in the Pacific Ocean feeding in immense numbers on living corals. On opening their bodies, their intestines were found full of impure chalk.

I have seen dried specimens of the white mud obtained from the sea-bottom, in the West Indies, that could scarcely be distinguished from chalk. Similar conditions to those that now obtain in the neighborhood of some of our coral islands may have assisted in forming the chalk-beds.

But how shall we account for the existence of flint, sometimes found in nodules in the heart of the chalk, but most frequently in layers between the chalk-beds, three or four feet apart?— flint, so hard, often black, so utterly different from chalk. When we break flints, we frequently find fossils in their interior: shells, corals, and echini, or sea-eggs, are very common, turning out of the flint, under the hammer, as a kernel does out of a hazel-nut. Nay, the solid flint itself, presenting no appearance of organic forms to the naked eye, on being chipped off in transparent splinters, and placed under the microscope, frequently reveals infusorial shells in great numbers. In a sea swarming with life, abounding with infusoria, sponges, corals, echini, shells, and fishes, calcareous or limy mud settled to the bottom for a vast period; and, as the various forms successively dropped into it, they were enveloped in this white mud, which subsequently hardened into chalk. In this mud were abundance of silicious or flinty shells, as well as limy ones, flinty infusoria, and spines of sponges and silica, deposited by thermal springs, whose waters held it in solution. These all mixed at first with the calcareous mud at the sea-bottom, became separated in time, and formed the nodules of flint, so common, especially in the upper chalk. But how was this done? "Birds of a feather flock together," says the old proverb; but, what is very singular, minerals in a fine state of subdivision have a similar tendency. Felspar and quartz, ground to a fine powder and mixed together, as in the English and French potteries, require to be ground over now and then to prevent the separation of the clay and silica. I have noticed, where heaps of tailings are collected at quartz-mills, in Colorado, after some time the sand and sulphuret of iron form parallel bands, distinctly separated, though originally all mingled together. The nodules of chert found in sub-carboniferous limestone, and the kidney-ore of iron found in the coal-measures, were probably formed in a similar way. Thus flinty shells, sponges containing flinty spiculæ, and spiny echini, or sea-urchins, formed the centres around which the flinty particles gathered out of the soft, white mud; and hence it is that we so frequently find them in a fossil state on breaking open the nodules of flint.

The seas were drained eventually as the land at the bottom was lifted up, and the hollows of the ocean in other places deepened; both causes from an early time operating simultaneously, and increasing the land-surface of the earth. Then rivers ran over the soft rock, scooping out channels for themselves; the waves of the ocean beat against the elevated portions in their neighborhood, and thus in time were formed the tall cliffs of chalk found on the coast of Kent and Sussex, and the sea-beaches covered with flint-pebbles that lie below them.

The chalk formation of Europe extends from the north of Ireland to the Crimea in the south of Russia, eleven hundred and forty miles in length; and from the south of Sweden to Bordeaux in France, eight hundred and forty miles in breadth. Formations older than the chalk separate the northern chalk region from a southern one, which is found in Spain, Italy, Greece, and other countries bordering on the Mediterranean.

In England, geologists find the following beds,—those of the *lower cretaceous*, sometimes called the "greensand group," consisting of beds of green sand, sandstones, limestones, and clays; and the *upper cretaceous*, composed of gault,—a stiff, dark-blue clay, abounding in well-preserved shells; beds of greensand, called the "upper greensand;" lower chalk, generally destitute of flints; and upper chalk, abounding in flints. The average thickness of the chalk in England is estimated at a thousand feet.

In America, true chalk is almost or entirely wanting: and yet the cretaceous formation covers a large portion of the continent; following the Atlantic border, from New York to Florida, along the Gulf of Mexico to Western Texas, and northward through New Mexico and Colorado, where it reaches a height of from six to seven thousand feet on the slope of the Rocky Mountains, and from there probably through to the Arctic Ocean.

The beds consist of layers of sand, frequently green; shells, loosely packed or cemented together; clays, sandstones, and limestones. In New Jersey, beds of green sand are very common in this formation. It consists of little green grains, its color due to the presence of silicate of iron, mixed with sand and lime, which is spread upon the lands to fertilize them, and is known by the name of *marl*. The beds abound with shells having a great resemblance to those found in the chalk of Europe.

In Western Texas, beds of cream-colored limestone are found, which is called by the people there *chimney-rock;* for they build their chimneys of it. It is an excellent building-material; for, when first taken from the quarry, it is soft enough to hew with an axe, and smoothe with a carpenter's plane, and hardens by exposure to the air. The time will come when beautiful cities will be built of it. Beneath this are softer beds, nearly as white as chalk, containing fossils in great abundance; and, where the streams have eroded them, the traveller can walk over what appears like the beach of an ancient ocean, strewn with shells, corals, and fish bones and teeth. On the Clear Fork of Trinity River, I have seen a large extent of surface covered with hard, flinty shells (*gryphœa*), lying so close together that grass could not grow between them. Asking a farmer in the vicinity where the shells came from, his answer was, that he supposed the Indians were fond of them, and brought them from the Gulf, which is not less than three hundred miles distant. Little did he dream that that spot was once the bed of an ocean in which those shells flourished as oysters do on our coasts to-day; that they became enveloped in mud, which hardened in time into rock, the silicious particles filling the pores of the shell, and making it of flinty hardness; that the ocean-bed was raised, the water drained off, and the torrents of ages, sweeping off the soft limestone, left the hard shells to tell the wonderful story.

In Colorado, the cretaceous formation consists of immense beds of limestone, and thinner beds of sandstone, both abounding with fossil remains. The various branches of the Platte and Arkansas have worn out numerous and frequently wide valleys in them.

In the cretaceous beds, we have the first of true palms, and of trees possessing a true bark. More than a hundred species of the last have been collected. Among them are the oak, beech, poplar, willow, alder, buttonwood, dogwood, tulip-tree, and sassafras. These have been found in New Jersey, Alabama, Nebraska, Kansas, New Mexico, and Vancouver's Island. In England, a few seaweeds and cone-bearing trees have been found; but few remains of land-plants have been discovered. In Colorado, there are coal-beds of this age; so that, in some places, swampy forests must have existed, as in the time of the coal-measures. Large beds of lignite are found on the continent of Europe in strata of this age.

Radiate animals are very abundant in the cretaceous beds; the most common being the echinus, which is found abundantly in the English flints, and still more abundantly in some of the limestones in Texas. I have seen fossil echinites in some streams there nearly as abundant as pebbles are in others. One species is called by the people there "the lone star of Texas," from the star-like markings on its upper surface. Fig. 32

Fig. 32.

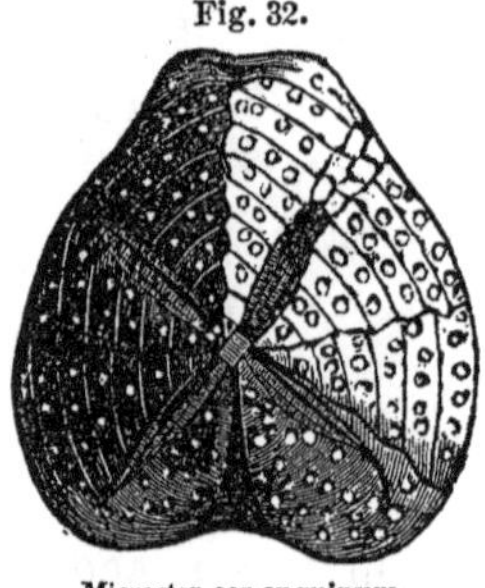

Micraster cor-anguinum.

Fig. 33.

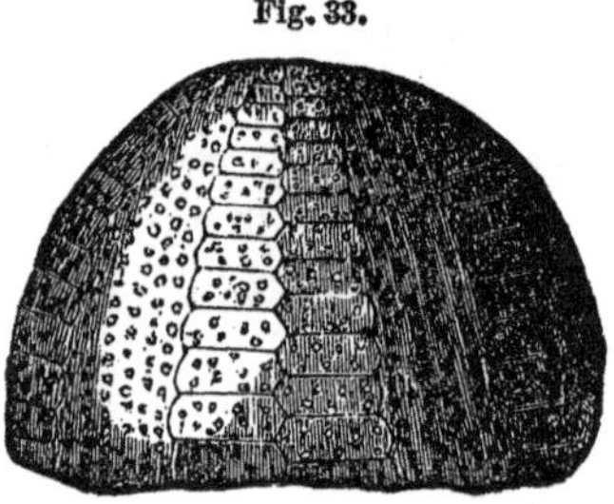

Ananchytes ovata.

represents one species from the chalk of England; and Fig. 33, another from the chalk of France. *Micraster* means "little star;" and *cor-anguinum*, "snake-heart;" its shape and its markings suggesting the name. *Ananchytes* means "not pressed;" and *ovata*, "egg-shaped;"

the shape again suggesting the name. Corals, sponges, star-fishes, and other zoöphytes, are frequently found in flint-nodules; the sponges often giving shape to the flints, their pores having been filled by silicious matter.

Fig. 34.

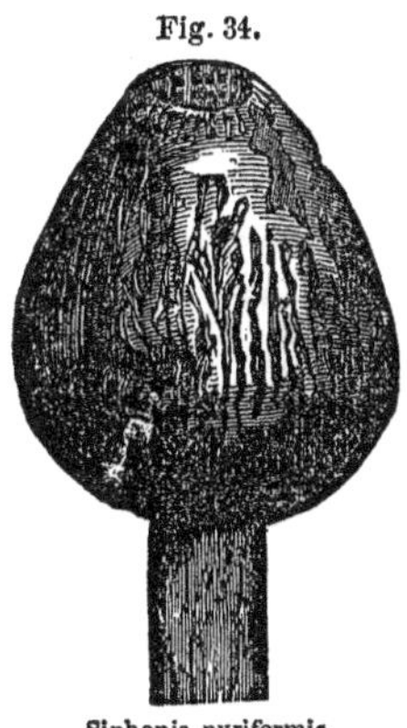

Siphonia pyriformis.

Fig. 35.

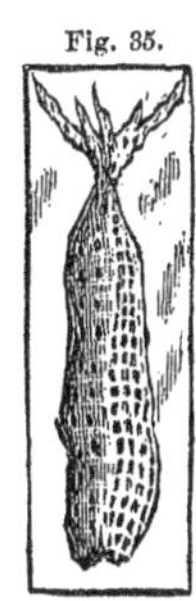

Veutriculite.

Fig. 34 represents a pear-shaped sponge from England; and Fig. 35, another cretaceous sponge, somewhat like a funnel: they are of solid flint.

Shells of many kinds are common in cretaceous rocks. More than a thousand species are known. Some grew to an immense size, and were of great beauty. Cephalopods, such as the ammonite and nautilus, abound. Several cephalopods existed which are apparently modified forms of the ammonite. The *baculite*, from the Latin *baculum*, "a walking-stick," resembles a straight ammonite. The *turrilite* (Fig. 36), from

Fig. 36.

Turrilite costatus.

Fig. 37.

Scaphites Ivanii.

the Latin *turris*, "a tower," may be described as an ammonite twisted in a tower-like form. The *scaphite* (Fig. 37), from the Latin *scapha*, "a skiff," and the *ancyloceras*, "curve horn" (Fig. 38), are ammonites with the shell partly uncoiled. The *crioceras*, "ram's horn"

(Fig. 39), is an ammonite, with a space between the whorls. I have seen the children in Texas playing with

Fig. 38.

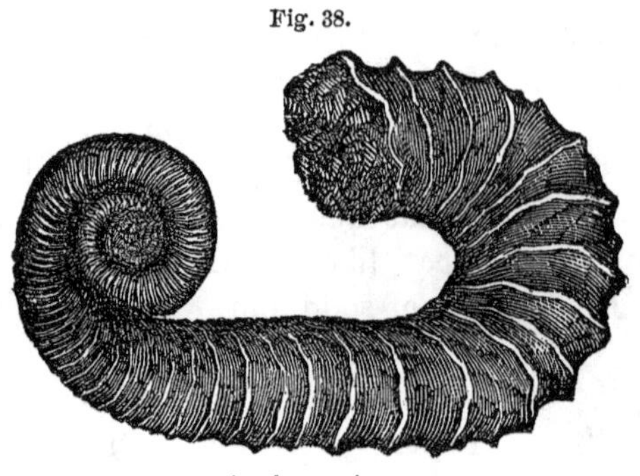

Ancyloceras gigas.

large ammonites, which they trundled to see which could make them go farthest, just as boys do with hoops. The baculite is very abundant in the cretaceous beds of Colorado, where it is styled a fossil-fish. It is sometimes found several feet in length. The cretaceous beaches must have presented a handsome appearance strewed with these enormous and beautiful shells. The ocean itself could not probably be better described than in the words of Percival: —

Fig. 39.

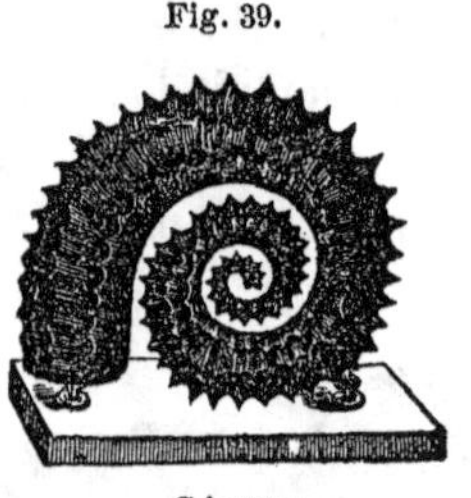

Crioceras.

> "Deep in the waves is a coral-grove,
> Where the purple mullet and gold-fish rove;
> Where the sea-flower spreads its leaves of blue,
> That never are wet with the falling dew,
> But in bright and changeful beauty shine
> Far down in the green and glassy brine.

The floor is of sand, like the mountain-drift;
And the pearl-shells spangle the flinty snow:
From coral-rocks the sea-plants lift
Their boughs where the tides and billows flow.
The water is calm and still below,
For the winds and the waves are absent there;
And the sands are bright as the stars that glow
In the motionless fields of the upper air.
There, with its waving blade of green,
The sea-flag waves through the silent water;
And the crimson leaf of the dulse is seen
To blush like a banner bathed in slaughter:
There, with a light and easy motion,
The fan-coral sweeps through the clear, deep sea;
And the yellow and scarlet tufts of ocean
Are bending like corn on the upland lea;
And life, in rare and beautiful forms,
Is sporting amid those bowers of stone;
While the wrathful Spirit of storms
Has made the top of the waves his own."

Speaking of some of the French cretaceous beds, D'Orbigny says, "It seems as if the sea had retired in order to show us, still intact, the submarine fauna of this period, such as it was when in life. There are here groups of polyps, echinoderms, and mollusks, which lived in union in animal colonies analogous to those which still exist in the coral reefs of the Antilles and Oceanica. In order that these groups should be preserved, it was necessary that they should be covered at once, and suddenly, by the sediment which is now, after being destroyed by the action of the atmosphere, revealing to us in their most secret details the nature of the ages which have passed."

The waters abounded with fish, some of which have been found in the chalk absolutely round and perfect as when they were alive. Some rocks are covered with the impressions of their scales, and others are largely made up of their teeth and bones. Near the Greenhorn River, one of the branches of the Arkansas, the sandstones, which alternate with limestones, and form terraces along the river-valleys in consequence of their hardness, are in some places literally paved with teeth; the most common belonging to species of the genera *ptychodus*, meaning "folded tooth," and *acrodus*, meaning "hump tooth," so called from the shape of the teeth. The waters abounded in sharks, though they do not appear to have attained the size that they did subsequently. We do not often find their bones, for they had cartilaginous or gristly skeletons; but portions of the shagreen or pimpled skin have been found in the chalk of England.

Several specimens of the *macropoma*, a fish from one to two feet in length, have been found in the chalk nearly perfect. Even the membranes of the stomach are preserved, and separate in flakes; and the ramification of the minute vessels is visible with a high magnifying power. With the remains of the macropoma have been found coprolites containing scales and bones of fishes, affording evidence of its predaceous character.

Of reptiles, the cretaceous formation in England has yielded twenty-four species. The ichthyosaurus and plesiosaurus occur, though rarely; and various fragments of the pterodactyle have been found. These three forms seem, however, to have died out with the cretaceous period; at least, no fragments have yet been found in the younger groups of rocks.

In this formation, the *mosasaurus*, or great reptile of the Meuse, a river in Holland, makes its appearance.

Its head (Fig. 40) was discovered in a quarry near Maestricht on the Meuse, by Dr. Hoffman, from whom it

Fig. 40.

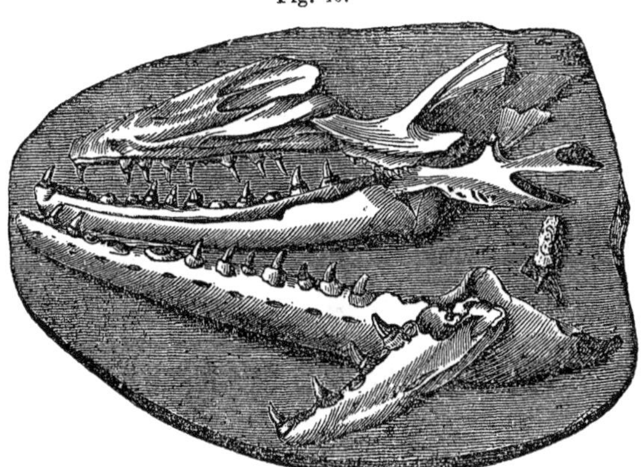

Mosasaurus Hoffmanni.

receives its specific name; and at first puzzled naturalists considerably. Some thought it a whale; and others, a crocodile. It is, however, a marine lizard, allied to the monitor of the Nile. Its head is five feet long, and its whole length is estimated at twenty-four feet. The teeth are directed backward, like the barb of a hook; and there is no doubt that it was a voracious monster.

Marine turtles were abundant; but the absence of all mammals is remarkable. As they existed previous to this, there is no doubt that they existed during this time; and, when the cretaceous beds of America shall have been examined as thoroughly as those of England have been, I have no doubt that this missing page in the volume will be found.

In the marl-pits of New Jersey, along with remains

of the saw-fish and sword-fish, have been found those of huge crocodiles and lizards. A thigh-bone of one herbivorous reptile, discovered near Haddonfield, measures nearly four feet in length. Another, a carnivorous reptile, supposed to have been about eighteen feet in length, was furnished with sharp, serrated, knife-like teeth, and had hind-legs nearly seven feet long, while its fore-legs were not more than one-third as long. This reptile must have widely differed from any of its modern relations, leaping with agility upon its prey. We have seen but a few of the grand hosts that have tenanted earth, sea, and air in the past ages.

Three species of birds have been discovered in this formation. One found in the chalk in Kent, England, resembles the albatross; another, found in Switzerland, is not unlike a swallow; and a third, found in the Cambridge greensand, is a bird about the size of a woodcock.

In Europe and the eastern part of this continent the difference between the life-forms of the cretaceous and tertiary periods is so great, that it has been said that no single species passes from one to the other. But along the base of the Rocky Mountains, and between the ranges, many forms are common to both; so that it is difficult to distinguish some of the bordering groups of rocks.

During the cretaceous period, the sea covered a large portion of the North-American continent. Commencing at New Jersey, the ocean occupied a strip of varied width from there westward, along the coast to Texas, including the whole of Florida; extended up the Mississippi to the mouth of the Ohio, and in a wide belt from Texas, in a north-western direction, in a line with the Rocky-Mountain range, probably to the Arctic Ocean.

The ocean laved both sides of the Sierra Madre, or main chain of the Rocky Mountains, the range being several miles narrower than at present; for we find rocks formed during that period occupying summits two and three thousand feet above the level of the plain, and seven or eight thousand above the sea-level. The Colorado Parks seem to have been inland seas, probably communicating, by what are now passes over the range, with the exterior ocean.

TERTIARY PERIOD.

AGE OF MAMMALS.

In the infancy of geology, all the rocks were simply divided into primary, secondary, and tertiary, or first, second, and third; but as the science advanced, and the earth's crust became better known, many other groups of rocks were found, and other terms became necessary. But, though the terms "primary" and "secondary" are seldom used by geologists now, "tertiary" has been retained, and is applied to those rocks which lie between the cretaceous formation and the drift.

The tertiary period has been divided by Lyell into three, — the eocene, miocene, and pliocene. Eocene is derived from the Greek *eos*, "dawn," and *kainos*, "recent," and signifies the dawn of the recent; for, at the time the name was given, about six per cent of the shells found in the eocene beds were believed to be identical with living species. Although we now know that this is not the case, the name is still appropriate: for the outlines of land and water, and the forms of life, begin to resemble those now existing; and we may in them see the prophecy of the present, as the first beams of morning herald the coming day. Miocene is from *meion*, "less," and *kainos*, "recent;" and pliocene from *pleion*, "more," and *kainos*, "recent;" the pliocene being more recent than the miocene, and the miocene less recent than the pliocene.

The tertiary period has been called the age of mammals.

Mammals are those animals that suckle their young. They made their appearance, as I have observed, before the tertiary period; but it was then they attained their largest size, and flourished in greatest abundance. The class *Mammalia* has been divided by geologists into various orders; those mammals being grouped together that have the greatest resemblance to each other. The most common arrangement makes twelve of these orders:—

First, *Bimana*, or two-handed mammals. The only animal belonging to this order is man.

Second, *Quadrumana*, or four-handed mammals. In this order, all apes, monkeys, and lemurs are placed. All the four limbs are used for grasping. They resemble man more nearly than any other animals.

Third, *Cheiroptera*, or hand-winged mammals. This includes bats, whose forward limbs are wing-like, enabling them to fly like birds.

Fourth, *Insectivora*, or insect-devourers. This includes shrews, moles, and hedgehogs, whose teeth are so arranged as to enable them most readily to crush the hard skins or shelly coverings of the insects that constitute their food.

Fifth, *Carnivora*, or flesh-eaters. This is a large group, containing all animals of the cat and dog kind, as well as weasels, bears, and other flesh-eating mammals. In them, sharp teeth pass each other like the blades of a pair of scissors; and thus the flesh on which they feed is easily divided.

Sixth, *Cetacea,* or whale-like mammals. These are fish-like in their appearance, and adapted for a life in the water. This order includes the whale, the grampus, the porpoise, and some others. It includes the largest mammals, and indeed the largest animals that exist.

Seventh, *Rodentia,* or gnawing mammals, such as rats, hares, beavers, &c. The animals of this order are generally of small size, but are very numerous. They have four front teeth, or incisors, that are very sharp, by which they readily gnaw nuts, roots, and even large trees.

Eighth, *Edentata,* or toothless mammals. All the animals belonging to this order are not, however, literally toothless, though some are; but all are destitute of front teeth. In this order are placed sloths, ant-eaters, and armadilloes.

Ninth, *Pachydermata,* thick-skinned mammals. The elephant, the rhinoceros, the horse, and the hog belong to this order. Most of the animals placed in this order have thick and naked skins. The largest land-mammals belong to this order.

Tenth, *Ruminantia,* the cud-chewers. All animals that chew their cud belong to this class, such as the ox, camel, sheep, and deer. This order contains the greatest number of animals useful to man.

Eleventh, *Marsupialia,* the pouched mammals. This includes those animals which have a pouch for their young, into which they are placed, and where they are cared for till they are strong enough to shift for themselves. The opossum and the kangaroo are of this kind.

Twelfth, *Monotremata.* Mammals with one excretory orifice, like a bird. This includes the ornithorhynchus and some other singular animals of New Holland.

These orders are again divided into families, these into genera, and the genera are lastly divided into species. Thus the wolf, the fox, and the jackal belong to the genus *Canis*, which is the Latin for "dog:" but the dog, as a species, is known as *Canis familiaris*, or the common dog; the wolf, as *Canis lupus;* and the jackal, as *Canis aureus.* The cat, the lion, the tiger, and the leopard belong to one genus,—*felis*, or "cat." The cat, is *Felis catus;* the lion, *Felis leo;* the tiger, *Felis tigris;* and the leopard, *Felis leopardus.* In this way, the scientific names for all animals are formed; the first word expressing the genus, and the last identifying the species.

When the differences between animals are slight, as between the cart-horse and the race-horse, the animals possessing them are said to be of different varieties; when they are greater, as between the horse and the ass, they are of different species. There is little doubt that varieties are undeveloped species, and species undeveloped genera; all lines of distinction drawn between them being merely artificial.

It is probable that animals belonging to all these orders, except the first, existed during the eocene period; though some of them were much better represented than others.

Under the city of London, and in the vale of the Thames generally, lies a bed of clay, known by the name of "London clay," which varies in thickness from two hundred to six hundred feet. On boring for water in the neighborhood of London, the workmen, on passing through this clay to the underlying chalk, obtain a copious supply often rising to the surface. At the Isle of Sheppy, at the mouth of the Thames, which is entirely composed of London clay, it is exposed in cliffs, some of which are two hundred feet high, to the action of the waves; so that the fossils which are contained in it are washed out in great abundance. Mr. Bowerbank collected here twenty-five thousand specimens of fossil fruits alone, belonging to six or seven hundred different species. None of them are now living; but they resemble tropical fruits, such as the date, cocoanut, areca, custard-apple, gourd, melon, coffee, scarlet-bean, pepper, and cotton-plant, and therefore indicate a climate such as now exists in Ceylon and the West-India Islands. Forests of spices grew, and aromatic shrubs gave their odor to the morning breeze; but we look in vain for beings sufficiently advanced to enjoy them. How few there are, even now, that fully enjoy what Nature bestows with so lavish a hand! Many species of shells and fish, fifty species of which have been described by Agassiz, indicate, in like manner, a warm climate.

Many extinct species of crabs and lobsters have been found in the London clay, also the remains of crocodiles, a serpent twenty feet long, a small vulture, a kind of king-fisher, a heron, a species of sea-gull, an opossum, and a bat.

That portion of England where the London clay is found which is called the "London basin" appears at that time to have been a gulf, into which fruits, seeds, and wood were probably drifted by currents; and, as Mantell says, "the existence of a group of spice-islands at no great distance seems necessary to account for so vast an accumulation of vegetable productions."

In France is a Paris basin, corresponding in some respects with the London basin, though the beds are more numerous, and composed generally of different material. They consist principally of sands, marls, limestones, and gypsum, or plaster. At Montmartre, close to Paris, quarries have been opened, and gypsum taken out for the manufacture of plaster-of-Paris. As the workmen dug out the plaster, they discovered bones in it, but supposed them to be the bones of existing animals, such as dogs, horses, and sheep; but the attention of Cuvier, the great French naturalist, being directed to them, his knowledge of comparative anatomy enabled him to see that these bones were different from all that he had previously observed. He went to the quarries, stimulated the workmen by presents, and secured all the bones that they could obtain. These he collected into a room, and then commenced the work of reconstructing the animals to which they had belonged. He says, "I at length found myself as if placed in a charnel-house, surrounded by mutilated fragments of many hundreds of skeletons of more than twenty kinds of animals, piled confusedly around me. The task assigned me was to restore them all to their original position. At the voice of comparative anatomy, every bone, and fragment of a bone, resumed its place. I cannot find words to express the pleasure I experienced in seeing, as I discovered one character, how all the consequences that I predicted from it were successively confirmed. The feet were found in accordance with the characters announced by the teeth; the teeth, in harmony with those indicated beforehand by the feet: the bones of the thighs and legs, and every connecting part of the extremities, were found set together precisely as I had arranged them before my conjectures were verified by the discovery of the parts entire."

Professor Ansted, using the language of Cuvier as far as possible, shows us how such re-creations are possible : —

"Thus, if the stomach of an animal is so organized as only to digest fresh animal food, its jaws must also be so contrived as to devour such prey, its claws to seize and tear it, its teeth to cut and divide it, the whole structure of its locomotive organs to pursue and obtain it, its organs of sense to perceive it from afar: and Nature must even have placed in its brain the necessary instinct to enable it to conceal itself, and to bring its victim within its toils ;" and this would give a certain shape to the skull, which the comparative anatomist could recognize.

"That the jaw may be enabled to seize the prey, there must be a certain shaped prominence for its articulation; a certain relation between the position of the resistance and that of the power with respect to that of the fulcrum ; a certain magnitude of the muscle that works the jaw, requiring corresponding dimensions of the pit in which that muscle is received, and of the convexity of the arch of bone beneath which it passes; while this arch must also possess a certain amount of strength to enable it to bear the strain of another muscle.

"That the animal may be able to carry off its prey, a certain degree of strength is necessary in the muscles which support the head: whence results a peculiar structure in the vertebræ to which these muscles are attached, and in the back of the skull where they are inserted.

"That the teeth may be adapted to tear flesh, they must be sharp, and they must be more or less so exactly according as they are likely to have more or less flesh to tear; while their bases must be strong in proportion to the quantity of bone and the magnitude of the bones they have to break. Every one of these circumstances, also, will have its effect on the development of all the parts which assist in moving the jaw.

"That the claws may be able to seize the prey, there must be a certain amount of flexibility in the toes, and of strength in the nails; and this requires a peculiar form of the bones, and a corresponding distribution of the muscles and tendons. The fore-arm must possess a certain facility in turning, whence also result certain forms of the bones of which it is made up; and these bones of the fore-arm, articulating to the humerus, cannot undergo change without corresponding changes taking place in this latter bone. The bones of the shoulder, also, require to have a certain degree of strength when the anterior extremities are to be used in seizing prey; and in this way, again, other special forms become involved."

Thus the bony structure of an animal corresponds exactly with the life that it must live; and a man familiar with the relation between them can sometimes, from a small fragment, determine the form of the perfect animal and its habits. Seeing the track of a cloven foot, we might be sure that it was made by a ruminating mammal; and this would give us a knowledge of its teeth, jaws, skull, vertebræ, pelvis, and body.

Some of the animals found in the Paris basin, Cuvier called *paleotheres*. *Palaios* is the Greek for "ancient;" and *therion*, "animal,"—"ancient animals;" and well do they deserve the name. Millions of years have passed since their bodies were floated down some milky stream, and deposited in the mud which formed the gypsum of Montmartre. They were of various sizes, from that of a hare to that of a horse. They more nearly resembled the South-American tapir than any other existing animal. They probably lived in rivers, and fed upon the plants that grew on their margins. Along with these have been found the remains of the wolf, fox, raccoon, dog, opossum, and squirrel; of birds, — the buzzard, owl, quail, woodcock, curlew, and pelican. Yet, though I call up familiar animals by these names, they do not convey to us the exact appearance of them; for, although they belong to the same genus as those animals, the species are all distinct. The dog, for instance, differed from the living dog as the wild cat differs from the tame, or as the ass differs from the horse.

Shells called nummulites (from *nummus*, the Latin for "a piece of money") are very common in the rocks of this period, and, being found only in tertiary rocks, are a very important guide to the geologist in determining the age of many beds in which they are contained.

"The nummulitic formation," says Lyell, "with its characteristic fossils, plays a far more conspicuous part than any other tertiary group in the solid framework of the earth's crust, whether in Europe, Asia, or Africa. It often attains a thickness of many thousand feet, and extends from the Alps to the Carpathians, and is in full force in the north of Africa; as, for example, in Algeria and Morocco. It has also been traced from Egypt, where it was largely quarried of old for the building of the Pyramids, into Asia Minor, and across Persia to the mouths of the Indus. This remarkable formation enters into the loftiest portions of the Alps, and has been found in Western Thibet sixteen thousand five hundred feet above the level of the sea." The eocene limestone of Suggsville, Ala., which forms hills three hundred feet high, is entirely composed of forms allied to them.

What a story these small, circular shells tell! It is evident, that, when they lived, the Alps, Pyrenees, Carpathians, and even the lofty Himalayas, lay beneath the ocean, or were but lifting their heads above it. What a difference there would be between a map of the world then and now! And yet, before this time, England had long existed, and been peopled by birds and beasts of various kinds.

The continent of North America appears to have been slowly lifted out of the water as one formation after another was laid down at the bottom of the ocean. In consequence of this, as we pass from the mountainous portions of the continent toward the sea, as a rule, we pass over rocks more and more recent, till we arrive at the coast. The formations thus extend in belts widening toward the west along the Atlantic States.

South of the cretaceous formation, along the Atlantic border, from Delaware and Maryland to Georgia and Alabama, we find an eocene belt, widening as we go south and west. The beds are composed of sand, marl, clay, and sandstones, and limestones abounding with fossils. At Claiborne, Ala., four hundred species of shells and remains of echini and fishes have been found. In Clarke County, in the same State, have been found immense bones belonging to a sea-monster called the *zeuglodon*, or yoke-tooth, from the yoke-like appearance of its teeth. Some of its vertebræ are so heavy, that they are a load for a man to carry; and its length was at least seventy feet. So numerous are they, that Lyell says, "I obtained evidence, during a short excursion, of so many localities of this fossil animal within a distance of ten miles, as to lead me to conclude that they must have belonged to at least forty distinct individuals."

It was at first supposed to be a lizard, and was named *basilosaurus*, or king of lizards; but it is now known as a mammal of the whale kind, and therefore a member of the order *cetacea*. It somewhat resembled the *manatus*, or sea-cow. Animals of the same genus have been found in miocene beds of France, Germany, and Malta.

Somewhat older than the beds containing the zeuglodon, though frequently regarded as more recent, is a deposit at Brandon, in Vermont, consisting of clay and gravel and (lying in these in irregular bodies) manganese, iron-ore, and lignite or brown coal. In the brown-coal bed have been found bushels of fossil fruits, most of them nuts. Among the fruits have been recognized some that resemble the cinnamon, the fig, and the grape.

The climate of Vermont during the eocene period was evidently warm, though, as its *flora*, or plants, indicate, not as much so as in places of corresponding latitude in Europe: thus early, that difference in climate which exists between the two continents made its appearance.

There was, doubtless, a teeming *fauna*, or assemblage of animals, feeding on the fruitful spice-forests of eocene Vermont; but the first fragment has yet to be discovered.

Miocene. — The beds of the miocene, or middle tertiary, are not found in England, but are well developed in Europe and America. The vegetation indicates a cooler climate than the preceding periods, though still warmer than exists in the same latitude at the present day. Tropical palms, bamboos, and laurels grew in Europe, and with them the maples, walnuts, beeches, elms, and oaks of a temperate clime.

In Nebraska, and the country of the Upper Missouri, is a remarkable region, called by the French *mauvaises terres*, or "bad lands." It consists of immense beds of clay which have been carved by running streams flowing through numerous winding channels, and causing the portions left in relief to assume the appearance of fantastic ruins. The region is thus described by David Dale Owen: —

"From the uniform, monotonous, open prairie, the traveller suddenly descends one or two hundred feet into a valley that looks as if it had sunk away from the surrounding world, leaving standing all over it thousands of abrupt, irregular, prismatic, and columnar masses, frequently capped with irregular pyramids, and stretching up to a height of from one to two hundred feet or more.

"So thickly are these natural towers studded over the surface of this extraordinary region, that the traveller threads his way through deep, confined, labyrinthine passages not unlike the narrow, irregular streets and lanes of some quaint old town of the European continent."

Fig. 41.

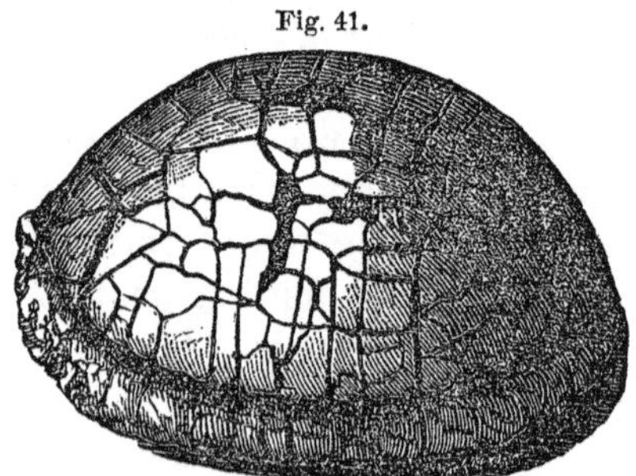

Testudo hemispherica.

Numerous fossils, once embedded in the clay, have been washed out, and obtained; among them, many turtles. Fig. 41, from this region, derives its name from its hemispherical shape. The weight of some specimens now lying there has been estimated at a ton. The tertiary period seems to have been eminently favorable for turtles, especially the early part of it: most deposits of that age contain remains of them. They basked on the logs that lay in the rivers; they floated on the tepid waters of the shallow lakes; they made paths along the sandy shores to their nests, where eggs innumerable were buried, from which new broods were continually raised.

Forty species of extinct mammals have been discovered in the beds of the *mauvaises terres*: eight of them carnivorous, or flesh-eaters, and related to the hyena, dog, and panther; and twenty-five herbivorous, or vegetable-feeders, and resembling the deer, hog, camel, and horse. The jaw of one paleothere (resembling those of the Paris basin, but larger) measures five feet along the range of the teeth. It was twice as large as a horse. One skeleton measures eighteen feet long, and nine high. One species, styled *archeotherium*, unites characters belonging to three orders: the molar teeth are constructed after the model of the hog; the canine teeth are like those of the bear; while the upper part of the skull and cheek-bones have the form and dimensions of the cat tribe.

The *oreodon*, another extinct form from these beds, has grinding teeth like the elk and deer, and canine teeth resembling the hog. They were animals, it is supposed, that lived on flesh and vegetables, and yet chewed the cud.

That portion of Nebraska must have been covered by a shallow and probably brackish lake during this time, having rivers pouring in turbid torrents, and sweeping down the remains of the various animals living where their waters ran: thus they became mingled with the remains of the turtles, the inhabitants of the lake.

Near the junction of White and Green Rivers, partly in Colorado, and partly in Utah, is an immense tertiary deposit, which I regard as miocene. It consists of a series of petroleum shales, a thousand feet in thickness, varying in color from that of cream to the blackness of cannel-coal. The shales abound in impressions of leaves and of various species of insects.

Fossil insects are rare, considering that at the present time insects are the most numerous of all animated forms. A few fragments have been obtained from the Devonian rocks of New Brunswick, and more numerous fragments from the carboniferous beds of England, Nova Scotia, and Illinois. At Solenhofen, in Bavaria, very per-

Fig 42.

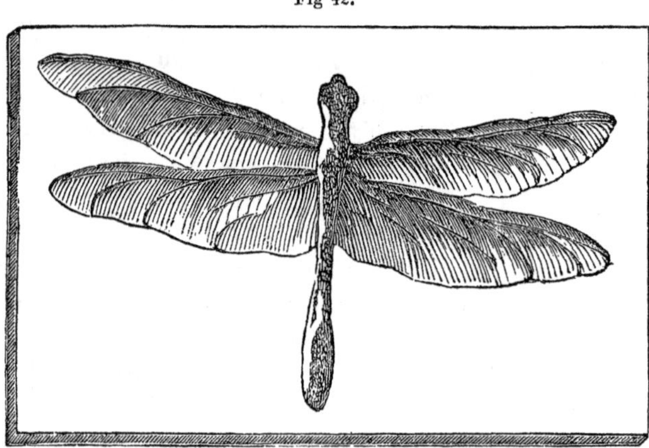

Æschna eximia.

fect specimens have been found; among them a beautiful dragon-fly (Fig. 42), which measures from tip to tip six inches. In the lower lias of England, several bands of limestone occur, which are crowded with the wing-cases of beetles, and occasionally perfect insects, and hence are known by the name of "insect limestones." But I think there is no part of the world known where fossil insects are as numerous and as perfect as in the petroleum shales of the White-river basin. The collection that I brought from there was placed in the hands of Mr. S. H. Scudder, Secretary of the Boston Natural-history Society, whose knowledge of this much-neglected branch of paleontology is probably greater than that of any other American. He says the collection consists of between sixty and seventy species of insects, representing nearly all the different orders. "About two-thirds of the species are flies, — some of them the perfect insect, others the maggot-like larvæ. The greater part of the beetles were quite small. There were three or four kinds of homoptera (allied to the tree-hoppers), ants of two different kinds of genera, and a poorly-preserved moth. Perhaps a minute thrips, belonging to a group which has never before been found fossil in any part of the world, is of the greatest interest. At the present day, these tiny and almost microscopic insects live among the petals of flowers; and one species is supposed by some entomologists to be injurious to the wheat." Mr. Scudder adds, "It is astonishing that an insect so delicate and insignificant in size can be so perfectly preserved on these stones. In the best specimens, the body is crushed and displaced, yet the wings remain uninjured; and every hair of their broad but microscopic fringe can be counted."

I brought the specimens from two localities, — one in Utah, near the mouth of White River; and the other about sixty miles up the stream, in Colorado. I have no doubt that the insect shales extend from the one locality to the other, and cover hundreds of square miles, in a country where the rocks are exposed by denudation in ten thousand places. Future explorers will find rich harvests in this highly interesting region.

Above the shales containing the insects are alternating beds of sandstone, shale, and conglomerate, not less than a thousand feet in thickness. In the conglomerate beds are rolled fragments of bones of large mammals, generally solid; and in the sandstones perfect turtles, having shells of great thickness and denseness. The country formed of these beds is one of the most remarkable on this continent. From the summit of a high ridge on the east, a tract of country containing five or six hundred square miles is distinctly visible. Over the whole surface is rock, bare rock, cut into ravines, canyons, gorges, and valleys; leaving in magnificent relief terrace upon terrace, pyramid beyond pyramid, rising to mountain-heights, and these pyramids ruled with delicate lines from base to summit, caused by the stratification of the shales of which they are composed; amphitheatres that would hold a million spectators; Titanic walls, castles, towers, pillars, statues, everywhere. It looks like some ruined city of the gods, blasted, bare, desolate, but grand beyond description. Originally an elevated country, composed of a number of beds of sandstone, conglomerate, and shale, of varying thickness and hardness, it has been worn down and cut out by rills, creeks, and streams, leaving this strange weird country to be the wonder of all generations.

During the time that these beds were being deposited, a large basin occupied by a lake, or, more probably, several connected lakes, covered Western Colorado, and extended into Utah and Arizona. Into them flowed numerous streams, carrying down fine mud, and at times petroleum in large quantities, probably flowing from numerous springs. The land was covered with trees resembling the oak, maple, and willow; though other forms that existed are now extinct. Insects by myriads hovered around the margins of the lakes, and their larvæ swarmed in the peaceful waters; turtles abounded; and aquatic pachyderms sported among the reeds, and bathed in the lakes, some feeding upon the fish, and others on the numerous plants that sprang from their muddy bottoms. The water was probably salt, or at least brackish; the present Salt Lake of Utah being a modern representative of these older and larger lakes, into which rivers poured mud, sand, and pebbles, sweeping down leaves, branches, and occasionally trunks of trees and skeletons, till the shales, sandstones, and conglomerates of the White-river basin were formed.

Fig. 43.

Terebratula grandis.

The miocene formation is well represented in the Atlantic States by beds of marl, clay, sand, limestone, sandstone, and lignite. At Martha's Vineyard in Massachusetts, and in a belt from New Jersey along the Atlantic border to South Carolina, miocene beds are found. They contain a great abundance of shells, of which from ten to forty per cent are of living species. Some of them are solid masses of broken shells. Fig. 43 represents one of the tenebratulæ of this time. This mollusk was furnished with a muscular arm, or pedicle, which it protruded through the hole at the end of the shell, and thus anchored to surrounding objects.

Professor Heer has discovered in the miocene beds of Northern Greenland the fir, poplar, beech, plane, hazel, and other plants that are now only to be found eight or nine hundred miles to the south. The most common tree appears to have been one allied to the red-wood of California, which sometimes attains a height of two hundred and seventy-five feet, and a diameter of twenty. The climate of Greenland must have greatly changed since such vegetation flourished there.

In the neighborhood of Œningen, near Constance, in Germany, are miocene beds of marls, limestone, and lignite or imperfect coal. The Rhine flows through them, exposing beds on both sides in cliffs from seven hundred to nine hundred feet in height. In these beds

Fig. 44.

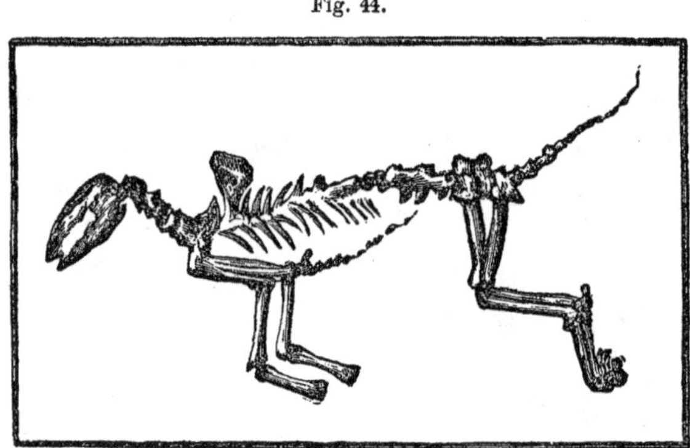

Galecynus Œningensis.

have been found leaves of the poplar, willow, maple, lin-den, and elm. The buckthorn has also been found, and what is supposed to be a species of wheat. In the same beds have been found insects, shells, fishes, turtles, tortoises, birds, and an animal resembling the common fox, *Galecynus Œningensis* (weasel-dog of Œningen), Fig 44. Owen gives it an intermediate position between the pole-cat and dog.

Amber has been found in the beds at Martha's Vineyard and at Brandon; but the greatest amount has been obtained on the shores of the Baltic, washed out of the lignite beds by the waves. Species of pines existed, from which gum, or resin, flowed; and, becoming fossilized, amber was the result. In flowing down the tree, insects, spiders, small crustaceans, and leaves were covered; and thus we find them preserved in the transparent amber. Upwards of eight hundred species of insects have been observed, and eight species of coniferous trees, or trees bearing cones like the pine, with "several cypresses, yews, junipers, oaks, poplars, beeches, &c.; altogether ninety-eight recognizable species of trees and shrubs. There are also some ferns, and numerous mosses, fungi, and liverworts."

Fish remains have been found in great abundance in miocene beds; among them teeth of sharks, larger than a man's hand. Sharks must have occasionally attained the size of large whales; and this indicates a great abundance of the small fish on which they fed.

At Monte Bolca, in Italy, is a hill composed of limy shales which abound in fishes in the most beautiful state of preservation; the scales, bones, fins, and even the muscular tissues, being frequently preserved. Several hundred species have been found there, and thousands of specimens carried off to enrich cabinets: still millions remain. "From the immense quantities," says Dr.

Mantell, "which occur in so limited an area, it seems probable that the limestone in which they are embedded was a limy mud erupted into the ocean by volcanic agency, and that the fishes were thus suffocated and surrounded by the calcareous mass." This is very likely; and such catastrophes must have taken place times innumerable. Life over limited areas destroyed; the relics of life deep buried at times, at others left to decompose, and leave no visible trace behind, again to be renewed, and again destroyed and renewed, — these to continue till the Earth shall attain her perfect condition, and peace and order reign everywhere.

At Œningen, already mentioned, in 1725 a remarkable skeleton was discovered (Fig. 45). It was examined by Scheuchzer, a Swiss naturalist, who supposed it to be the skeleton of a man, and gave a description of it in the Philosophical Transactions of London. In 1731, he published a pamphlet in reference to it, and entitled it "Man's Testimony to the Deluge." He says, "It is certain that this is the half, or nearly so, of the skeleton of a man; that the substance even of the bones, and, what is more, of the flesh, and of parts still softer than the flesh, are there incorporated in the stone; in a word, it is one of the rarest relics which we have of that cursed race which was buried under the waters." He

Fig. 45.

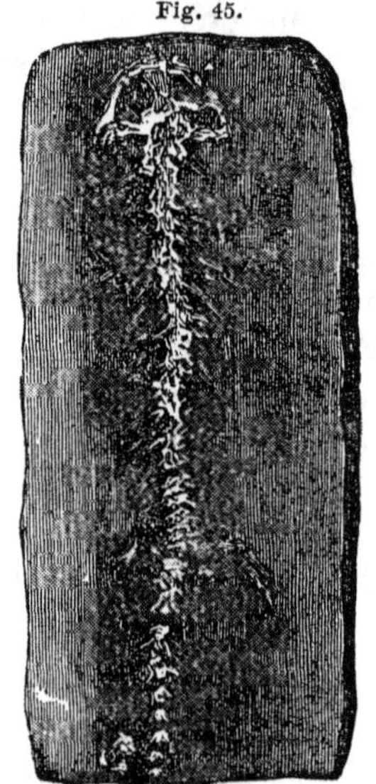

Andrias Scheuchzeri.

even says that there were remains of the brain, the roots of the nose, and some vestiges of the liver. Camper, a celebrated naturalist of Holland, examined it, and saw at once that it was not a man, but supposed it to be a lizard. The matter was decided by Cuvier, who pronounced it a large salamander; and thus perished the "witness of the Deluge."

One of the most remarkable mammals of this time was the *deinotherium*, or fearful beast. The name is derived from the Greek *deinos*, "fearful or terrible," and *therion*, "beast, or animal." It was of enormous size and peculiar appearance, but, being a vegetable feeder, was probably harmless. Its remains are found in the Valley of the Rhine, near Darmstadt, and in the Valleys of the Jura Mountains, and particularly at Eppelsheim, in Hesse Darmstadt, where thirty other species of fossil mammals have been found. An entire head of the deinotherium was found there, which measured four feet in length, and three feet in breadth; while the whole length of the animal was calculated by Cuvier to be eighteen feet. The head was furnished with a proboscis, and two large tusks in the lower jaw which were curved downwards. Dr. Buckland supposed that it lived in the water, floating its enormous body upon the surface, and using its tusks as pickaxes to grub up the roots on which it fed, and as anchors attached to the bank to support its head while it slept. They seem to be linking forms between the sea mammalia and the land.

At the same place have been found bones of the *mastodon*, which in various species during the remainder of the tertiary period, and even as recently as the human period, roamed over America, Europe, and Asia. This animal resembled the elephant, but was larger; his body being longer, and his limbs more massive. The teeth also were different, having conical eminences covered with enamel: hence its name, from *mastos*, "nipple," and *odous*, "tooth."

Six mastodon skeletons were found in Warren County, New Jersey, by a farmer who was digging mud from a small pond. Within the ribs of one, just where the stomach must have been, seven bushels of vegetable matter were extracted : some of it, upon examination, proved to be shoots of the cedar. Within the ribs of mastodons found in other localities have been found masses of small branches, leaves, grass, and reeds in a half-bruised state; but these animals must have existed, geologically speaking, within a recent period.

Tusks have been found measuring from twelve to fourteen feet in length. Imagine them! — white, smooth, curved, massive, a black trunk swinging between them, or stretched to an overhanging branch; back of these an enormous head, long, high, and arched, furnished with teeth weighing from ten to twelve pounds; with flapping ears, large as a blacksmith's leather apron, attached to a stack of flesh covered with a thick skin, and supported on legs like huge pillars. Multiply it by twenty, and you have a herd of mighty mastodons, such as wandered through the woods of Missouri and Kansas before the prairies began to be. Now they feed at the edge of the forest, tearing down the branches with their trunks, ever and anon trumpeting to their comrades in a neighboring forest, whose answer sounds like the mutterings of distant thunder. Away they go crashing through the woods, the young ones waddling after them, tearing up good-sized trees by the roots as they pass, for sport, down to the lake, and bathe their bare, black sides, while from their trunks they send up fountains of water. In the distance, see that pack of wild dogs hunting like wolves! What are they pursuing? A herd of wild horses. Hear them snort with fear, and see them toss their manes, as they go flying like the wind, with these gaunt dogs in full pursuit! Tapir-like animals are bathing in the rivers, and sloths hang from the branches of far-spreading trees; grasses cover large beaver-meadows; while monkeys, the most intelligent onlookers, chatter to each other on the surrounding trees, and leap from bough to bough.

The fossil horse has been found in various tertiary beds, ranging from the miocene to the drift, in France, Germany, Malta, India, and America from New England to Patagonia. Those found in the miocene generally differ from the existing horse in the distribution of the enamel of the tooth, and in the possession of two small digits and hoofs, which dangled by the side of the large one. They appear to have been transitional forms between the paleotheres of the eocene and the horse of the present.

Though the horse was common on the American continent during the tertiary, yet, when the Spaniards first came here, there was no animal of the horse kind in America. The Indians had never seen one; and they supposed, when they saw the Spaniards on horseback, that horse and man were all one animal.

One by one, old forms die out. The trilobites of the early times disappeared, one generation after another flourishing and decaying, till the last vanished; the ammonites of the oölitic and cretaceous periods, the great reptiles, and the flying pterodactyles, flourished, lingered, and died; and so the mastodons, camels, and horses of America. As the leaves of evergreens constantly drop, and new ones supplant them: so with the great tree of life; the species as leaves dropped, and new ones gradually supplanted them, so that it is only by looking back over immense periods that we can see the great changes thus produced.

There were many mammalian forms existing during the miocene period, whose living appearance would startle us considerably. Between the River Sutlej and the Ganges, in India, is a tertiary deposit, consisting of beds of conglomerate, sandstone, and loam, spread over the flanks of a range of hills belonging to the sub-Himalayan Mountains, known as the Sewalik Hills. In these have been found paleotheres, elephants, mastodons, giraffes, rhinoceroses, camels, antelopes, and monkeys, birds of the ostrich kind, crocodiles, and tortoises.

The most remarkable, perhaps, among this wonderful assemblage of animals, is the *sivatherium*, deriving its name from the Indian god Siva. It was as large as an elephant, and probably taller; and seems to have been a link between it and the ruminants, or cud-chewers, — an elephantine stag. It had four horns, and was furnished with a proboscis. One pair of the horns was short, and placed forward above the eyes; the other pair broad and long, like those of the elk, rising above the ears. It had small eyes and great lips.

Among various reptilian remains, including turtles and crocodiles, some of which cannot be distinguished from species now living in India, are the bones, and portions of the shell, of a gigantic tortoise, not less than twenty feet in length.

The tertiary formation has a remarkable development in Colorado. I estimate its thickness, from the top of the Parahlamoosh Range (which lies west of Middle Park, and is composed of tertiary lavas, probably of miocene age) to the junction of White and Green Rivers, at ten thousand feet. Included in this are coal-measures, containing many beds of coal, underclays, sandstones, shales, and thin limestones, having a total thickness of twenty-four hundred feet. The lavas of the Parahlamoosh, with their alternating beds of sandstone, cannot, I think, be less than four thousand feet in thickness. Since their deposition, the range must have been greatly elevated, as it is now thirteen thousand feet above the sea-level; and yet its beds are apparently horizontal. The lavas that formed them were evidently poured out on the bottom of a lake, or inland sea, covering the sand that lay there, and in some cases penetrating it to a depth of ten or twelve feet. Between some of the outpourings, sand was swept down over the lava, covering it, in places, to a depth of more than a hundred feet; then another layer was poured out over this; and so on for thousands of feet.

In the Middle Park, Colorado, are a series of beds, which I have called baculite beds, from the abundance of baculites contained in them. They are associated with gasteropods and conchs, that resemble tertiary fossils much more than they do cretaceous, although the baculite is characteristic of the upper chalk in England and on the continent of Europe. These beds belong, I think, to the eocene period. They are succeeded by shaly sandstones four hundred feet thick, abounding with the impressions of the scales of cycloidal fishes; that is, fishes whose scales resemble those of the salmon, the trout, and most of our modern fishes: but although the scales are so numerous, and some of them an inch in diameter, not a tooth was discoverable.

Above these sandstones are other sandstones and conglomerates, about four hundred feet in thickness, abounding in fragments of silicified wood and occasional bones, probably of recent miocene age; and above these are beds of lava, about two hundred feet in thickness, containing moss-agates, and geodes of chalcedony, which, where the beds have been disintegrated, strew the ground.

At the foot of the Rocky Mountains, on the east, is a deposit of tertiary coal, or lignite, which burns readily, but will not coke. At Golden City, twelve miles west of Denver, one bed has been opened which has a thickness of from ten to twelve feet. Owing to upheaval, this bed, with its accompanying shales and sandstones, stands nearly vertical. About twenty miles north of Denver, on Coal Creek, there are several exposures of coal alternating with beds of iron-ore and sandstones, resembling very much the beds of the Welsh coal-measures from which most of the iron of Great Britain is produced.

It is not difficult, with the discoveries already made in Colorado, to call up the country as it existed on the eastern side of the mountains about the close of the miocene period, after the deposition of the lignite beds; the mountain-chain then probably less elevated, destitute of its snowy crown, and clad with coniferous trees to its summit everywhere. Immediately west of Golden City were immense forests of lordly trees, the growth of ages; some palms; some resinous and gum-bearing trees of strange aspect, on which fed lordly mastodons, while sloths hung from their branches by their long claws as they ate the foliage.

Below was a long and wide lake, covering the spot where Golden City and Denver now stand, and stretching north and south for an immense distance. From the character of the surrounding country, this lake must have been clear as crystal; and we may see the fish that bathe in its pellucid waters, the turtles that crawl at the bottom or lazily float on the top, while alligators fight with each other, and seize the fish that constitute their prey. Mountain-goats, with their long, recurved horns, look down from the rocky crag that juts over the lake; and the wading-birds feed upon fish and frogs, its innumerable tenants.

The remains of palms and resinous trees are very abundant along the base of the mountains on the eastern side for a great distance. Hot-springs charged with silica poured out their waters, which percolated through the sand-beds containing the remains of trees washed down by the mountain-torrents; and the face of the country, in some localities, is absolutely covered with fossil wood thus produced: for the land became the theatre of widespread and intense volcanic action, as indicated by trap deposits in Middle Park, near Golden City, on the eastern side of the range, and outlying masses, probably once connected from there to the Huerfano River, one of the branches of the Arkansas. These were probably produced by one of the last upheavals of the Rocky-mountain chain; the drainage of the lakes following, and the production of innumerable hot-springs, whose effects can be traced over so large a district on both sides of the range.

Two species of monkeys have been discovered in the miocene of Southern France. They resemble the genus hylobates, or long-armed apes of India. Of three species found in the Sewalik Hills, two resemble the present solemn ape of India, and the third the common Indian monkey; but all were of larger size than existing species. At Pikermi, in Greece, M. Gaudry has recently discovered the remains of fifty-one species of animals, all of which are extinct. Among them were twenty-five monkeys belonging to one species. The animals found are somewhat African in type, and show that Greece was intimately connected with Africa during miocene times.

Pliocene. — With hasty steps we approach the period of the present; and at every step changes are manifested which are of great interest to the observer. During the tertiary period, the Rocky Mountains were gradually uplifted several thousand feet, attaining, probably, a greater height than they at present possess. Out of the water have been heaved the Andes, the backbone of South America, the Alps, and the Himalayas, the mighty crests of Europe and Asia; lakes have been drained; immense plains wrested from the waters; and now the land-surface of the globe becomes very similar in outline to that with which we are familiar. The climate is cooler; the palms that flourished so abundantly in Europe disappear, and are succeeded by the vegetation of a more temperate clime. The shells found abundantly in many of the beds are similar to those of our present seas, being from seventy to ninety per cent modern.

Fig. 46.

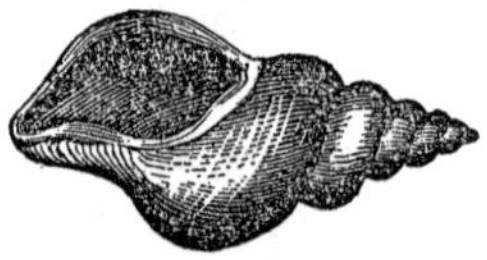

Fusus antiquus.

In Suffolk, England, there is a well-marked group of rocks belonging to this formation, known by the name of the "Suffolk Crag." One portion of it, called the Coralline Crag, is almost entirely made up of coral; though shells, sponges, and echini are distributed through it. One very character istic shell of the upper crag is the *Fusus antiquus* (Fig. 46), which is associated with shells belonging to genera now existing in temperate zones. Those which are peculiar to tropical climates are either absent or of small size. About four hundred and fifty species of shells have been found in the Suffolk crag, together with the remains of sharks, and bones of four species of cetaceans, or whale-like animals.

It is a remarkable fact, that all, or nearly all, the species of reptiles, birds, and beasts of the tertiary are extinct. The elephants of that time were mastodons and mammoths. The horses of that period had differently-formed teeth, and some of them differently-shaped feet. Fig. 47 represents the tooth of the horse of the pliocene, found at Suffolk, England. The oxen were generally larger, and their horns and teeth differed from those of the present time.

Fig. 47.

Equus fossilis.

Slowly, from age to age, animals appear to have been modified, changing in their structure as the conditions of existence changed around them. Those animals with which we are familiar are but a few of a host innumerable belonging to that grand life-procession that has marched through the ages. They are like islands in the ocean, — solitary representatives of perished multitudes; fragments, time-worn as those are wave-worn; in their turn to disappear, and be succeeded by others arising slowly and imperceptibly into being.

While the mastodon and mammoth were flourishing in North America, Europe, and Asia, animals equalling them in bulk, though differing widely in organization, existed in South America.

During a dry season, a countryman hunting for cattle discovered on the banks of the River Salado, south of Buenos Ayres, what appeared to be the trunk of a tree. Throwing his lasso over it, with the aid of his brother Guachos he dragged it on to the bank. It proved to be an enormous bone, five feet in diameter, — the pelvis of a *megatherium*, or monstrous beast, as it has been very properly named. Other bones were brought up at the same time, and among them several vertebræ. "To the Guachos, the pelvis luckily appeared to be useless: turn it which way they would, they all agreed that it did not make half so comfortable a seat as a bullock's head, the arm-chair of the pampas." Some time after, the bones were sent as curiosities to the owner of the land on which they were found: and here Sir Woodbine Parish discovered them, dug out many others, and had them conveyed to England; and it is from them that the casts of the megatherium have been made which are now in Boston, Cambridge, Amherst, and other localities.

Its length was eighteen feet; its height at the withers, seven feet; its pelvis, — the largest bone ever found belonging to a land-animal, — five feet broad; its fore-foot was a yard long, and twelve inches wide; its toes terminated by large and powerful claws of great length; its thigh-bone was nearly three times as broad as that of the largest elephant, its circumference at the largest part being three feet two inches. When clothed with flesh, the tail must have been more than six feet around.

At first sight, this seems to be one of the most monstrous, clumsy, and ill-formed creatures that ever existed; and yet, when its structure is fully understood, we see use, harmony, and perfect adaptability, everywhere. Let us first examine its teeth, that we may learn the kind of food on which so large an animal subsisted. It did not live on flesh: the teeth are broad, for grinding and chewing; not sharp, for cutting. It lived, then, on vegetable food; but what kind? Not on grass; for it has no front teeth, and on a prairie it would have starved to death. Not on bushes; for such a monster as this would have required half an acre of them for a breakfast, and it was too large and clumsy to travel fast or far. What then? It must have subsisted on trees; and, regarding this as its diet, how appropriate the whole organization appears! The pampas, now covered with grass, were at that time clad with immense forests; and rapid locomotion was altogether unnecessary to the animal that fed upon their trees. Such an animal could not climb them, and hang suspended from the branches, as the living sloth of South America does, which it most nearly resembles in structure: but this was unnecessary; for here was strength that could bring the branches and trees to it. Sitting on its massive hind-legs, supported by its powerful tail, forming a secure tripod, it reaches up to the branches, and, laying hold of them with its long fore-arms and strong claws, brings them within reach of its head: this head is so strong and substantial, and contains so many holes for the passage of nerves and blood-vessels, that it doubtless supported a proboscis, probably flat. Without stirring from its position, then, the megatherium could pull down branches or young trees, so as to be reached by the head, and ground by the teeth; distant ones could be reached by the proboscis, so that its ponderous body need not be moved from its position; and, even when the proboscis could not reach a far-off limb, it would seem, from the shape of the vertebræ of the neck, which fit into each other like one cup of a set into another, that the neck could be elongated, and the tree stripped, before it became necessary to move.

The remains of the megatherium have been found in all parts of the pampas for a distance of eight hundred miles. The carcasses of this and of other animals were floated down the streams, and sank in their muddy beds, and have thus been preserved.

With the megatherium have been found the bones of another gigantic sloth, — the *mylodon*, or "mill-tooth," as its name signifies. The length of the skeleton is eleven feet, including the tail. It appears to resemble the sloth more closely than does the megatherium.

Sometimes it appears that trees fell upon these animals, and injured them. One skeleton of a mylodon was found with the outer part of the skull broken in two places, one of which was entirely and the other partially healed. Both the megatherium and mylodon had skulls with outer and inner plates, separated by an unusual thickness of air-cells, so that the outer one might be broken, and the animal still live and recover.

Two other large sloths existed at the same time, in the same locality, — the *megalonyx* (*megale*, "great;" *onux*, "claw") and the *scelidotherium* (*scelidos*, the "thigh;" *therion*, "beast"), so named from the great breadth of its thigh-bone. Remains of species of the mylodon and megalonyx have been found in North America, though they are most abundant in South America; and remains of the mastodon have been found in South America, though less abundantly than in North America.

There is a remarkable correspondence between the extinct fossil mammals of recent tertiary deposits and existing mammals in the same countries. In that country alone in which the sloth is found have fossil sloths been discovered. Europe, Asia, and Africa have furnished no fossil animal resembling the sloth, armadillo, or platyrhine monkey; that is, a monkey with a broad division between the nostrils: but in South America, where alone sloths and armadilloes exist to-day, we find their gigantic predecessors; and all the fossil monkeys are platyrhine, as are all the living ones. One of the extinct armadilloes is called the *glyptodon* (Fig. 48) (*gluptos*, "sculptured;" *odous*, "tooth"), from its molar teeth being marked on the sides with grooves. It was an animal with a shell something like a turtle, and formed of hexagonal plates. It measured eleven feet, following the curve of the back; and its armor was six feet eight inches in length, and nine feet across. It has been estimated that this carapace, or body-armor, weighed more than four thousand pounds. The structure of the legs and feet was well adapted to support this enormous weight; and at the same time the hind ones were allowed such free motion as was necessary in digging and scratching.

Fig. 48.

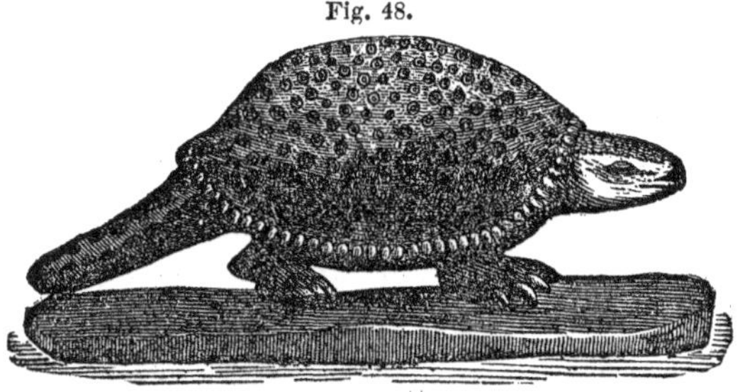

Glyptodon clavipes.

"Not a relic," says Owen, "of an elephant, a rhinoceros, a hippopotamus, a bison, or a hyena, has yet been detected in the caves, or more recent tertiary deposits, of South America. Not a single animal belonging to these genera is found living on that continent to-day; but in the Old World, where they do exist, we find the fossil remains of their predecessors."

The Australian tertiary deposits have in like manner yielded large extinct mammals resembling the living forms of that country. All the existing indigenous mammals of that country are marsupial, except the *dingo*, or "Australian dog." The fossil mammals which are found in caves and recent deposits are exclusively those of marsupials; though the species differ from existing ones, and the animals are generally much larger.

The living wombats, phalangers, potoroos, and kangaroos were preceded by extinct species of the same genera; some of the kangaroos being of great size. One of these gigantic kangaroos, that must have nearly equalled the hippopotamus in size, has been named, by Owen, *diprotodon* (animal having two front teeth). The skull, which is now in the British Museum, measures three feet in length. The hind limbs were much shorter and stronger, and the front limbs were longer and larger, than those of the living kangaroo. The skull (Fig. 49), with part of the skeleton, was discovered in a pliocene deposit on the Plains of Darling Downs, Australia.

In the same formation have been found the remains of a carnivorous marsupial rivalling the lion in size. One extinct wombat was as large as a tiger. Australia seems to have become less favorable for monstrous marsupials; and they were eventually succeeded by the smaller and feebler specimens that abound there at the present time.

When New Zealand was first discovered, no mammals were found upon it, with the exception of a rat; but it contained many birds, — the most remarkable, a small

Fig. 49.

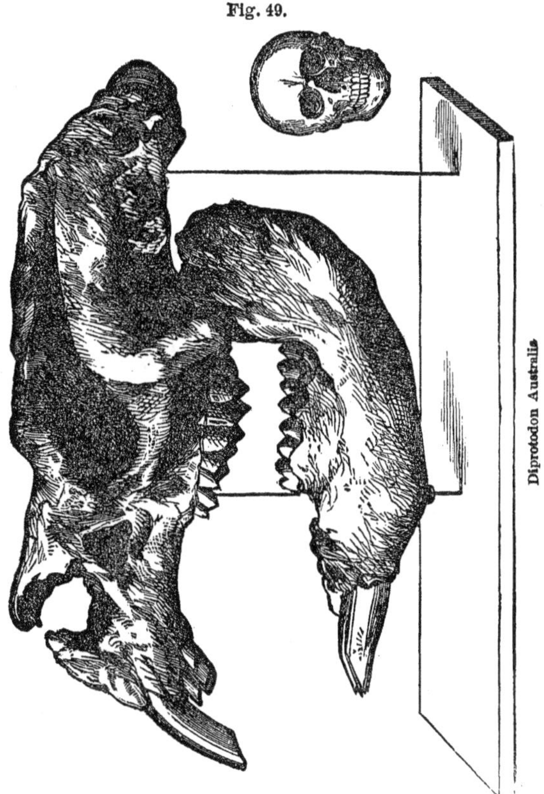

Diprotodon Australis

struthious bird, that is, a bird of the ostrich family, called the *apteryx* (wingless), on account of its very small rudimental wings. In a specimen whose body measured nineteen inches, the wings, stripped of the feathers, were only an inch and a half long; ending in a hard, horny claw three inches long. Several species of birds similar to the apteryx formerly existed in New Zealand, ranging in height from three to ten feet. "A small portion of a bone," says Mrs. Somerville, "which from its dimensions appeared to belong rather to a quadruped of the size of an ox than to a bird, was submitted to Mr. Owen. He boldly pronounced it, from its structure, to belong to a bird of the ostrich kind, — a decision that was soon abundantly confirmed by the discovery, not only of the bones of the bird, but of its eggs." This bird has been called the *dinornis* (huge bird); and well it may. In one specimen, the leg-bone measured upwards of a yard in length; and, in another, the toe-bones almost rival those of the elephant. Several species of the dinornis have been found; and large wingless birds appear to have abounded in the large islands of New Zealand at no very distant period. The bones have been found well preserved, and in some cases marked by edged tools, and mixed with human bones in the ash-heaps of ancient feasts: so that the extinction of these birds may be fairly attributed to savage man, who found them good eating, and easy to kill, as they could not fly; while man was not sufficiently civilized to domesticate and thus perpetuate them.

In Madagascar, portions of a very large bird, the *epiornis* (tall bird), have been discovered, and entire eggs measuring from thirteen to fourteen inches in diameter. One of these eggs, it is computed, would contain six ostrich eggs, or one hundred and forty-eight hen's eggs.

The ostrich family is at present represented in Australia by the *emu;* in the islands of the Indian archipelago, by the *cassowary,* — next to the ostrich, the largest of living birds; in Africa, by the ostrich, which ranges from Barbary to the Cape of Good Hope, its wide and uninhabitable deserts affording it security; in America, by the *rhea,* which is found from Bolivia to Paraguay; in New Zealand, by three species of *apteryx.* No bird of this family is found either on the continent of Europe or Asia, probably on account of its destruction by man, as the early Dutch navigators destroyed the *dodo,* once so abundant in the Mauritius, and as the *solitaire,* another wingless bird, has become extinct in the Island of Rodriguez and Bourbon, where it once existed in great numbers.

In the pliocene deposits of Montpellier, France, remains of a monkey occur; and in brick earth of this age, in Essex, England, a fossil jaw and tooth of the *macacus,* or bonnet-ape, have been found. Many extinct species of monkeys were discovered by Dr. Lund in South America, — all of them platyrhine, but much larger than living ones, many of them twice as large.

LECTURE V.

As our planet never twice occupies the same point in space as it journeys round the sun, so it has never twice occupied the same position in its progressive geologic movement, nor been twice in the same condition, from its nebulous birth to the present time. Its course has been progressive; yet its pathway has not been perfectly straight, nor the rate of its march invariably uniform. As, in the advancing spring-time, we are checked in our anticipations of fine weather by bitter blasts as cold as winter's breath, so has it been in the progress of the world. There have been times when it seemed to be marching backward rather than forward; yet out of all, as the year advances to smiling May, the world has marched on to a brighter day.

GLACIAL PERIOD.

The next page in the world's history brings us to one of those backward-looking times when a spectator might have justly despaired of the world's future. This period is known as the drift, or glacial period; and is marked by beds of sand, gravel, clay, and bowlders from the size of a man's head to that of a meeting-house. There is one in Whittingham, Mass., whose length is forty feet, its horizontal circumference a hundred and twenty-five feet, and its estimated weight three thousand four hundred tons. I saw one in Maine, east of Bangor, that was twenty-three feet high, and seventy feet round, which, from its position and appearance, must have travelled a considerable distance. These peculiar beds and rocks, lying near the surface or on it, have attracted attention from the earliest times. Whence came they? The answer to this question seemed at first to be extremely easy. Men supposed they saw in them evidences of a universal flood, the waters of which, sweeping over the world, heaped up these various beds, and transported the bowlders. With our increased knowledge, it is impossible to accept such a theory of their origin. Drift-beds or bowlders are not found farther south than about the latitude of 39°, or the line of Washington, Cincinnati, St. Louis, and Kansas City, except in high-mountain regions. In Tennessee and Alabama, there are no drift-beds or bowlders; neither are there any in Louisiana and Texas: and it is not until we go as far south of the equator as we find these beds north of the equator that we discover similar deposits. A universal deluge would make beds as universal; though then they would differ very widely from drift-beds: but the fact that these are confined to certain districts disproves the early and common theory of their origin. In fact, there are no geologists, in the proper sense of that term, who believe that there ever was a time, since man became an inhabitant of the globe, when it was covered with water to the tops of the highest mountains. It is simply impossible. There is not water enough about the planet to do it; and all the animals now living could never have been preserved by man.

On this subject, the Rev. Dr. Pye Smith has the following sensible remarks: "All land-animals having their geographical regions, to which their constitutional natures are congenial, — many of them being unable to live in any other situation, — we cannot represent to ourselves the idea of their being brought into one small spot from the polar regions, the torrid zone, and all the other climates of Asia, Africa, Europe, and America, Australia, and the thousand of islands, their preservation and provision, and the final disposal of them, without bringing up the idea of miracles more stupendous than any that are recorded in Scripture. The great decisive miracle of Christianity, the resurrection of the Lord Jesus, sinks down before it."

Hugh Miller shows us that no such deluge has taken place since the tertiary period. "In various parts of the world, such as Auvergne in Central France, and along the flanks of Ætna, there are cones of long-extinct or long-slumbering volcanoes, which, though at least triple the antiquity of the Noachian deluge, and though composed of the ordinary incoherent materials, exhibit no marks of denudation."

All floods since the Silurian period must, in the nature of things, have been partial; and of these there must have been multitudes, many of them since man has existed. In consequence, nearly all nations having a written or traditional history give us some account of a flood or floods. The inhabitants of Otaheite relate that their island was destroyed by the sea, and but one man and woman were saved. The Greeks had a tradition, thus set in poetry by Ovid: —

> "Impetuous rain descends:
> Nor from his patrimonial heaven alone
> Is Jove content to pour his vengeance down;

But from his brother of the seas he craves
To help him with auxiliary waves.
Then with his mace the monarch struck the ground:
With inward trembling, Earth received the wound;
And rising streams a ready passage found.
Now earth and seas are in confusion lost, —
A world of waters, and without a coast.
A mountain of tremendous height there stands
Betwixt the Athenian and Bœotian lands:
Parnassus is its name, whose forky rise
Mounts through the clouds, and mates the lofty skies.
High on the summit of this dubious cliff,
Deucalion, wafting, moored his little skiff:
He with his wife were only left behind
Of perished man, — they twain of human kind.
The most upright of mortal men was he;
The most serene and holy woman she."

When they came to land, the question arose in the minds of this worthy couple, "How can the world be repeopled?" seeing that they were too old to expect progeny of their own. Seeking counsel from the gods, Deucalion was advised to throw the bones of his mother over his head. This he understood to mean stones, — the bones of our mother-earth. On doing so, to his great astonishment and delight they were changed into men. The old lady's were, of course, transformed into women; and thus the world became inhabited as of old.

These various traditions, doubtless, had their rise in various local floods. The bursting of lake-barriers; irruptions of the sea, caused by a country's sinking below its level; long-continued rains; upheavings of the sea-bottom by earthquakes,—these and other causes have laid large tracts of inhabited land under water, and swept off great numbers of human beings; the survivors, owing to their ignorance of the earth's size, supposing the world to be destroyed. But the deposits formed by such floods must have been small, and they give us no clew to the origin of the glacial beds.

There is one appearance connected with these beds that may lead us to a rational theory of their origin. On digging below them, down to what miners call the *bed rock*, we find this rock to be covered with parallel scratches, or furrows, sometimes called striæ (from scratches as fine as the point of a pin could make to large furrows a foot broad and two or three inches deep), or else they are smooth and polished. Such appearances are very common in the neighborhood of Boston, at Roxbury, Dover, and Dorchester: they may be observed in the cuttings on all the railroad lines of Massachusetts; in all the mountain-regions of New England; everywhere in Canada; on the shores of Lakes Ontario, Erie, and Superior, — passing, in some cases, below the water-level, especially in Superior. I have seen them on the sandstones of the Western Reserve, in Ohio; on the limestone at Dayton, in the same State; in the bed of a stream near the Kansas River; and on the Rocky Mountains above Empire City, in Colorado, and on the Parahlamoosh Range west of the Middle Park, at a height of more than twelve thousand feet above the sea-level.

What agent could do such work as this? When we have discovered it, we probably have a clew to the agent concerned in making the drift-deposits; for where we find the one, we find the other. The rocks found in the drift are also scratched or polished. These scratches may be observed on the sides of bowlders all over New England, but are most easily seen on fragments of limestone in Western drift-deposits. We know of but one t ing that could do this; and that is ice. In fact, ice is engaged in doing just such work at the present time. At the equator, snow does not melt that falls on mountains which are more than sixteen thousand feet above the sea-level: three miles above the hottest of summer weather is winter; and, in the temperate zone, the distance is much less. In the Swiss Alps, the snow does not melt at a height of nine thousand feet, but, as it falls, packs into ice, which accumulates till it forms masses, which move slowly down the side of the mountain in the form of rivers of ice. There are fifteen hundred square miles of ice in the Alpine range, from eighty to six hundred feet thick. These rivers of ice, or glaciers, move at various rates, according to the nature of the ground over which they pass, and the condition of the weather: from eight to twelve inches a day is an ordinary rate of advance. Some of the glaciers of the Alps are as much as twenty miles long, in places a mile wide, and have a thickness of many hundred feet. As these immense bodies of ice slowly but irresistibly move down the mountain-side, they break off masses of the rock over which they pass; and these become embedded in the bottom, and act as so many groovers, scoring the rocks over which they pass in parallel lines; while that ice which contains none of these, acts as a polisher of the rocky surface beneath. Thus Mr. Charles Martins says, "If we penetrate between the soil and the bottom of the glacier, taking advantage of the icy caverns which sometimes open at its edge or extremities, we creep over a bed of pebbles and fine sand. If we raise this bed, we find the rock beneath levelled, polished, ground down by friction, and covered with rectilinear stripes, resembling sometimes a small furrow, more frequently deep scratches, perfectly straight, as if they had been engraved by the aid of a burin, or even a very fine needle. If we examine the rocks on the side of a glacier, we find the same stripes engraved on them where they have been in contact with the congealed mass." Just such appearances I have seen in a thousand places over the Northern States and British Provinces, from Cape Breton to beyond the Rocky Mountains; and have no doubt they were made in a somewhat similar manner.

Where the temperature of a country is so low that snow cannot melt, the whole face of it becomes covered with snow, which consolidates into ice, forming one grand mass; and this moves toward the sea (the land sloping in that direction), until, reaching it, it breaks off in immense bodies, which go floating off as icebergs, carrying frequently rocks upon their surface, which, when they melt, are strewed on the floor of the ocean. This is the case with Western Greenland, as described by Dr. Rink, quoted by Lyell.

"In that country, the land may be divided into two regions,—the inland and the outskirts. The inland, which is eight hundred miles from west to east, and of much greater length from north to south, is a vast unknown continent, buried under one continuous and colossal mass of permanent ice, which is always moving seaward; but a small proportion only of it in an easterly direction, since nearly the whole descends towards Baffin's Bay." On reaching the coast, it presents a perpendicular wall of ice two thousand feet high, and slopes gradually towards the interior, to unknown heights, as far as the eye can reach.

"Although all the ice is moving seaward, the greatest quantity is discharged at the heads of certain large friths, usually about four miles wide. Through these the ice is now protruded in huge blocks, several miles wide, and from one thousand to one thousand five hundred feet in thickness. When these masses reach the friths, they do not melt or break up into fragments, but continue their course in a solid form under the salt water, grating along the rocky bottom, which they must polish and score, at depths of hundreds, and even more than a thousand feet. At length, when there is water enough to float them, huge portions, having broken off, fill Baffin's Bay with icebergs of a size exceeding any which could be produced by ordinary valley glaciers. Stones, sand, and mud are sometimes included in these bergs which float down Baffin's Bay."

There is good reason for believing that North America during the drift period was in a similar condition. An immense sheet of ice, in places thousands of feet in thickness, covered Canada, the British Provinces, nearly the whole of New England, and a large portion of the United States directly west; this body of ice moving toward the ocean in the most direct line that the elevations of the land would allow. Its general direction was toward the south, where it could melt, and room be found for the accumulating mass; transporting masses of rock, in some cases, for hundreds of miles. Bowlders of granite in Central Ohio and Indiana have been carried from Canada. I have seen a block of jasper conglomerate at Grand Rapids, in Michigan, and others near Delphi, in Indiana, that had been transported from the northern shore of Lake Superior. The northern shores of Long Island are strewn with bowlders of red sandstone and granite from Connecticut. Over thousands of square miles lay the great ice-field, rising in terrace beyond terrace to the north, slowly moving to the south. Like an enormous rasp, or file, it rounded the mountains on its march; filled up, by the *detritus* worn from the rocks, enormous cavities; and, on reaching the sea, pushed far into its waters, and floated off in icebergs innumerable, which, melting, strewed the ocean-bottom with sand, gravel, and bowlders.

I suppose any country from which water would flow into the sea, if cold enough to become covered with an icy sheet, would have slope enough for the ice to move to the ocean in glaciers. In some places, the force has been so great as to push the ice up considerable slopes.

This cold period appears to have come on rapidly. Down dropped the dome of everlasting cold: the seas congealed, rivers were stayed in their course, the great lakes covered, and buried deep as the Atlantic depths, life in its various forms was destroyed, and Winter swayed his icy sceptre over a wide realm once so summer-like and fair.

In Europe, a somewhat similar condition of things existed. All Switzerland, and a considerable portion of Spain and Italy, have been subjected to glacial action. Even as far south as Spain and Corsica, vast plains of ice existed. From Norway to Great Britain, an immense glacier extended, by which large bowlders were transported across the German Ocean. Traces of ancient glaciers have been discovered in all Northern Europe, Russia, Prussia, and in the Carpathian and Caucasus Mountains. In Russia, bowlders have been identified with ledges more than eight hundred miles to the north. Scotland and Wales sent out glaciers from their icy mountains; and the ocean to the south must have been crowded with icebergs.

It is remarkable that Southern Patagonia and Terra

del Fuego in South America, and the Falkland Islands lying east of Southern Patagonia, show similar signs of glacial action; drift-beds, bowlders, and striated or scratched rock-surfaces, being common. Drift-deposits have been traced as high as 41° south latitude. More remarkable is the fact, that on the western coast of Patagonia, in the latitude of 47°, which corresponds with that of Central France, a glacier exists even now and part of it at the sea-level. Capt. King says its length is fifteen miles, while it has a breadth in one place of seven miles. The deep sounds along the coast, from there south, are all furnished with glaciers. Charles Darwin says, on a still night, the cracking and groaning of the great moving masses may be distinctly heard.

Let us look at Nova Scotia, New Brunswick, and New England, as they once were, — a continental desert of snow and ice in the interior of the country, and the ocean fringed with a gigantic ice-wall for thousands of miles; icebergs high as the proudest steeple; crystal islands floating off to the south; birds in multitudes hovering over them like clouds. The sea has fish innumerable, and troops of walruses that feed upon them; whales are spouting, and white bears swimming from one ice-floe to another. Possibly man, fur-clad, armed with a stone spear, killed the seal that slept on the sea-margin, and even attacked the walrus and the bear. Icebergs, detached from the pushing glaciers, float off into the sea, and, melting, drop their rocky burdens.

"Mr. Scoresby counted in view, at one time, as many as five hundred icebergs drifting along in latitude 69° and 70° north; these bergs rising above the surface to a height varying from one to two hundred feet, and measuring from a few yards to a mile in circumference at the waterline; and as the mass of ice below the water is known to be always seven or eight times greater than that which appears above it, and the whole is loaded with fragments of rocks, the effect of the gradual melting of such masses must inevitably be to strew the bed of the sea with a vast quantity of gravel and erratic blocks." Mr. Scoresby calculated that some of them carried one hundred thousand tons of earth and stones. Many large bowlders now on dry land were probably dropped on the sea-bottom in this way during the drift period.

The snow at last melted; the mighty mountains of adamantine ice dissolved, and rivers rushed from their bases, sweeping away loose sediment, and changing its appearance as first left by the now retreating glaciers. Up sprang the grasses from the long-buried and ice-preserved seeds; and the trees again put forth their leaves, and their branches bowed to the breeze. The long winter was past, and a glorious spring came with the promise of a blooming summer.

But what could have made it so cold during this time? Nobody knows. Many theories have been formed to account for it; but none of them seems satisfactory. Some suppose that there are cold and hot regions in space; that as the motion of our planet around the sun gives us spring, summer, fall, and winter, so the motion of our sun around a grand central sun may produce seasons for the solar system. The drift period, then, was the winter: we are living in the spring, and may anticipate the summer. Very comfortable doctrine truly, especially in cold weather, but, like some other comfortable doctrines, has little or no basis. Astronomers know nothing of cold or hot regions in space; and, until some facts are produced to favor it, that theory, or rather hypothesis, may be dismissed.

It has been suggested by several, that, if the land was sufficiently elevated above the ocean during this period, all the phenomena connected with it must have taken place; and this, no doubt, is true. If the European continent should be elevated three thousand feet, most of its surface would be covered with a glacial sheet, overlaid by perpetual snow. If the North-American continent should be lifted up eight thousand feet above its present level, snow would fall upon its surface as at present, but, over a large portion of that surface, would fail to melt. In time, an immense glacier would cover the whole of the northern portion, extending as far south as the chain of lakes, and, slowly pushing out southward, would transport blocks from north to south; round the hills and mountains over which it passed; grind down the sandstone rocks to sand, and the shaly and slaty rocks to clay; and produce just such beds and appearances as we are familiar with. But what should elevate the continent thus, — still more the European continent in a similar manner, and during the same period? It is difficult to believe that elevations so extensive and so great could have taken place during any one period; and we are still left in the dark in reference to the cause of the intense cold of the glacial time.

It has been suggested that the inclination of the earth's axis to the plane of its orbit, which is 23° 28′, and is the cause of our present seasons, may have been so different during the glacial period as to make the frigid zone coincide with those portions of the earth where glacial action occurred during the drift period. This is, perhaps, the most reasonable supposition. Dr. Winslow remarks in his "Cooling Globe," "Suddenly transplant Greenland from its present connection with the bottom of the Atlantic to the south of Australia, what would be the geological and the geographical changes upon the surface of the whole globe? As small as Greenland is in superficial extent, the earth would instantly feel the disturbance of its equilibrium; and the inclination of its axis to the plane of its orbit would become sensibly affected. The physical results and changes in distribution of organic life that would necessarily follow would be universal in character and extent. Similar consequences would naturally follow the slow elevation or submergence of continents." Since all the planets the inclinations of whose axes are known differ from the earth in the amount of their inclination, it is not unreasonable to suppose that the earth's inclination was once different, and that it has been constantly changing to suit the changing condition of its surface. This would necessitate climatal changes to correspond, and may have produced the phenomena of the drift.

Geology is young; and there are many problems yet to be solved, and difficulties to be conquered, that our children will delight to grapple with and overcome.

The drift-beds are interesting to us from the fossils that they contain. During the early part of the drift period, when the icy covering began to extend southward, animals of many kinds, some now extinct, inhabited regions that the ice subsequently covered; and hence we find in the beds of that period their remains. But few have been found in North America; but in Great Britain and the continent of Europe they are numerous. From them we learn, that, during the drift period, two species of bats existed in Europe, — the great bat and the horse-shoe bat, both now living. Bears were numerous: three kinds at least have been determined. One of these was the common brown bear of Europe, a second smaller, and the third the great cave bear, which, according to Cuvier, sometimes attained the size of a large horse: its remains are numerous in England, France, Belgium, and Germany. In one cave at Gaylenreuth, it is said that fragments of eight hundred of them have been discovered.

A great cave tiger, twice as large as the living tiger, roamed over England and Europe; and with this a smaller tiger, a leopard, a wild-cat, and other animals of the cat kind.

In Great Britain, Europe, and Asia, the remains of a remarkable carnivorous animal, called a *machairodus*, have been discovered. This animal was about the size of a tiger; and from the size and shape of the canine teeth, which resembled a sword, must have been a destructive creature.

At the same time, beavers abounded, of much larger size than the North-American beaver; and an animal allied to it, but still larger, has likewise been found. Fig. 50 represents the skull of a beaver-like animal, one-fifth the natural size. It was found near Clyde, N.Y. This species is the largest of the *rodentia* ever discovered. The mammoth appears to have been very abundant. It is supposed that upwards of two thousand grinders have been dredged up by fishermen off the little village of

Fig. 50.

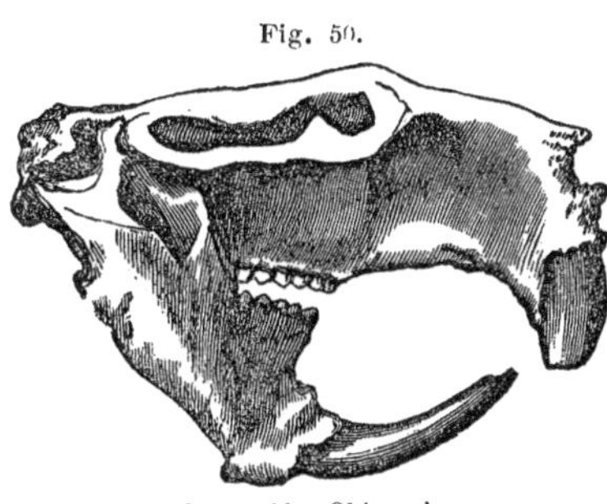

Castoroides Ohioensis.

Happisburg in Norfolk, England. The size to which they occasionally attained is indicated by Fig. 51, which

Fig. 51.

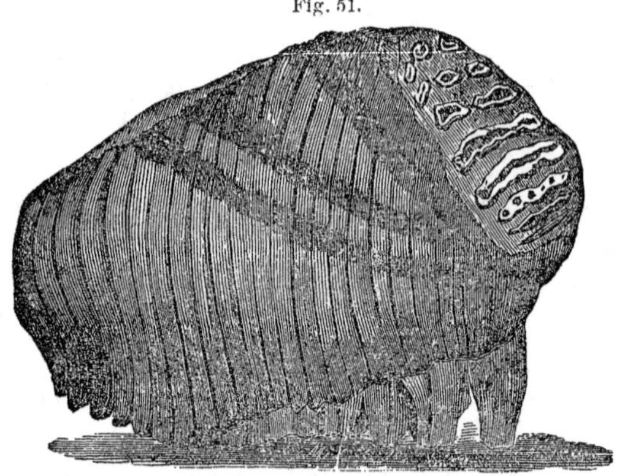

Elephas primigenius.

is one-fifth the natural size. The original, now in the Ward Museum, Rochester, weighs fourteen pounds.

But Northern Asia seems to have been the home of the mammoth. There is scarcely a river in which their remains have not been found; and even the islands in the icy sea north of Siberia abound with them. Belling, a Russian, who visited them, says of one, "All the island nearest to the mainland, which is about thirty-six leagues in length, except three or four small rocky mountains, is a mixture of sand and ice; so that when the thaw sets in, and its banks begin to fall, many mammoth bones are found. All the isle is formed of the bones of this extraordinary animal, of the horns and skulls of buffaloes or an animal which resembles them, and of some rhinoceroses' horns." Most of the ivory used by the civilized world is obtained from the mammoth

17

tusks of Siberia. In the year 1844, sixteen thousand pounds of teeth and tusks of mammoths were obtained, and sold in St. Petersburg.

In 1799, a Tungusian fisherman observed in an icy cliff at the mouth of the Lena an odd-shaped block which attracted his attention. Two years afterward, it was considerably exposed; and he saw the whole side and the tusks of a mammoth. In 1804, the season being warmer than usual, the body of the mammoth became disengaged from the ice, and fell on the beach; when the fisherman removed the tusks, and sold them. Two years afterward, Mr. Adams, who was on a mission from St. Petersburg to China, heard of the discovery, and visited the spot. The people of the neighborhood had cut off the flesh to feed their dogs, and wild beasts had also mangled it: the skeleton, however, was nearly entire. "The head was covered with a dry skin; one of the ears, well preserved, was furnished with a tuft, or mane; the balls of the eyes were distinguishable; the brain still occupied the cranium, but was dried up; the top of the neck was furnished with a flowing mane; the skin was covered with tufts of black hairs and reddish wool. More than thirty pounds of hair and wool were collected, which the white bears had taken from it, and trampled in the soil. The animal was a male: its tusks were more than nine feet long, and its head and tusks weighed more than four hundred pounds." Mr. Adams repurchased the tusks, carried the animal to St. Petersburg, and sold it to the Emperor of Russia for six thousand dollars.

The mastodon still existed. No extinct mammal with which we are acquainted had a wider range. Its remains have been found on every continent, and from the tropics to the polar seas. Of the hippopotamus there were two species, — one larger, and the other smaller, than that now inhabiting the Nile.

The remains of a two-horned rhinoceros have been found in Europe and Asia: its body was covered with hair, and destitute of those rough scales seen on the skin of the African rhinoceros. One of these, preserved in frozen sand, was found on the banks of the Viloui, in Siberia, in 1771; the very flesh still remaining on the bones, and the hairy skin covering it. This and the mammoth were evidently fitted for a colder climate than that in which the living species of the elephant and rhinoceros are now found; though neither could exist at present where their remains have been found, owing to the extreme rigor of the climate. It has been suggested, since the rivers run in Siberia from south to north, that their bodies may have been floated many hundred miles from a warmer country to the spot where they became embedded in ice. The bulky carcass of a mammoth or rhinoceros would float a long way before disintegrating by putrefaction. This occurring in the Lena towards winter, when there is ice at its mouth, it might be frozen in, and preserved for ages. This explanation is, however, I think, inadequate to account for the facts.

In the drift have likewise been found two species of the horse (one the size of the common horse, and the other that of the zebra), two species of camels, a gigantic musk-ox, a giraffe, deer of various kinds, and the great Irish elk.

The remains of the great Irish elk, *Megaceros Hibernicus* (great horn of Ireland), Fig. 52, are found abundantly in peat-bogs and marl-pits in Ireland, so that it is not considered a curiosity. It is found also in the Isle of Man, England, France, Germany, and Italy.

Fig. 52.

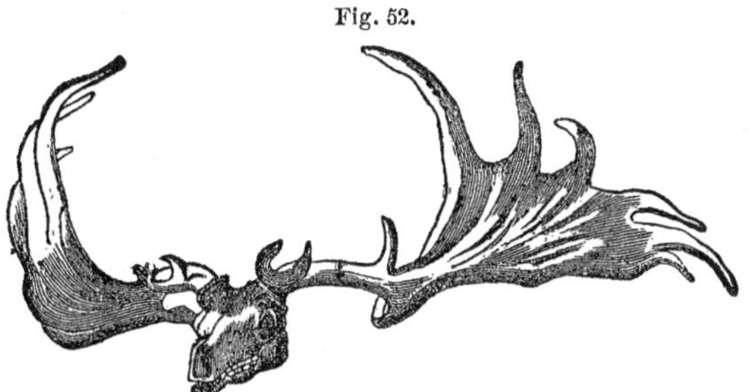

Megaceros Hibernicus.

Ireland was, perhaps, where they lived most recently, as their remains are found there in better preservation, and in greater abundance.

One perfect skeleton was found there, which is now in the museum of the Royal Dublin Society. From the tip of one horn to the tip of the other, it measures twelve feet, less two inches ; and from the ground to the highest point of the horn, nearly ten feet and a half. The weight of the skull of one, without the lower jaw, was five pounds and a quarter; while the same skull with lower jaw and antlers was eighty-seven pounds.

Man probably existed coeval with this enormous deer. In Germany, there were found in the same drain several urns and stone hatchets, and a head of this extinct elk. In Ireland, a human body was found, under eleven feet of peat, in good preservation, and clad in garments made, apparently, of the hair of this animal. There exists a rib in Dublin, which had evidently, during the lifetime of the animal, been pierced by some sharp instrument, apparently a stone arrow, as there was such a mark upon it as that would have made.

Several species of oxen were contemporaneous with the Irish elk, some of them of gigantic size. One, *Bos primigenius* (primitive ox), the skull of which is represented by Fig. 53, is said to have existed in Switzerland

Fig. 53.

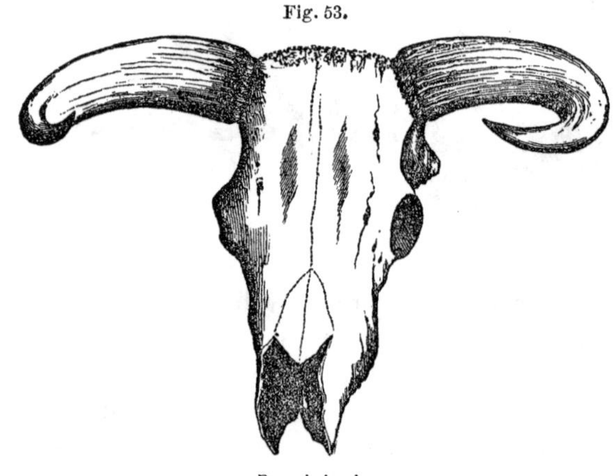

Bos primigenius.

as recently as the sixteenth century. One gigantic ox was nearly as large as an elephant.

But more important than all these, in the more recent drift and in caves, accompanying extinct animals, have been found the works and the remains of *man*.

Considerable attention was turned to the bones frequently found on the floors of caves, by the publication of a volume by Dr. Buckland on the Kirkdale Cave, discovered at Kirkdale, in Yorkshire, in 1820. "The bottom of the cave," says the doctor, "on first removing the mud, was found to be strewed all over, like a dog-kennel, from one end to the other, with hundreds of teeth and bones, or rather broken and splintered frag ments of bones, of the elephant, bear, hyena, rhinoceros, and hippopotamus. Those of the larger animals were found even in the inmost and smaller recesses. Scarcely a bone has escaped fracture. On some of the bones marks may be traced, which, on applying one to the other, appear exactly to fit the form of the canine teeth of the hyenas found in the cave." Mr. Gibson alone collected more than three hundred canine teeth of the hyena from this cave.

This appears to have been a den of hyenas for a long time. Into the woods they sallied, and dragged in the carcasses and bones of large animals which they found, and small animals which they overpowered and killed; and in the inmost recesses of the cave they lay and munched them. Dr. Buckland visited the Zoölogical Gardens in London; and, on presenting bones to hyenas there, he noticed that the marks made upon them by their teeth resembled those found on the bones discovered in the cave. Some of the bones of the cave hyena show that it occasionally attained the size of a Bengal tiger.

In 1840, a gentleman of the name of M'Enery, residing near Torquay, in Devonshire, England, found in a cave about a mile from the town, called Kent's Hole, human bones and flint knives among a great variety of extinct species, — such as the mammoth, two-horned rhinoceros, cave bear, and hyena, all from under a crust of stalagmite; and reposing upon it was the head of a wolf. Dr. Buckland visited the spot, and, on insufficient grounds, came to the conclusion that the remains of man and his implements were not as old as the bones of extinct animals that accompanied them; for Mr. Godwin Austen declared, in 1842, that he had obtained from the same cave, from undisturbed loam or clay under stalagmite, works of man, mingled with the remains of extinct animals; and that all these must have been introduced "before the stalagmite flooring had been formed."

In 1858, a new cave was discovered at Brixham, near Torquay; and it was resolved to examine it carefully and thoroughly. The Royal Society defrayed part of the expenses; and the work was carried on by the direction of Mr. Pengelly, a gentleman of much expeperience in searching caves, and Mr. Prestwich; Dr. Falconer taking an active part in examining the proceedings from time to time.

They found first a layer of stalagmite, from one to fifteen inches in thickness, in which some bones were found. Below this was loam or bone earth, from one foot to fifteen feet in thickness. The floor of the cave was found to be composed of gravel, that had a thickness of more than twenty feet. In the bone earth were found the remains of the mammoth, the two-horned rhinoceros, the cave bear, cave hyena, cave lion, and reindeer.

No human bones were found, but a number of flint knives, some of them even in the lowest gravel; but well-formed ones were taken from the bone earth, and one of the most perfect at a depth of thirteen feet in the bone earth, close to the left hind-leg of a cave bear. "Every bone," says Lyell, "was in its natural place: and it appears evident "that the bear lived after the flint tools were manufactured; or, in other words, that man in this district preceded the cave bear."

Long before this, discoveries of a somewhat similar kind had been made in the river-beds of France by

M. Boucher de Perthes. His first discoveries were made at Abbeville, on the River Somme, in 1841, and consisted of numerous flint implements, associated with the remains of the deer, horse, mammoth, rhinoceros, elephant, hyena, tiger, and hippopotamus. The implements consisted of flakes of flint, apparently intended for knives or arrow-heads; pointed implements from four to nine inches long, used for spear or lance heads; and oval implements, with a cutting edge all round.

M. Perthes published a volume in 1846, in which the implements were figured, and information given regarding their discovery; but, being too far in advance of his contemporaries, his book was disregarded, except by a few of his friends. A translation published in Liverpool met with the same fate.

In 1854, Dr. Rigollot, an eminent physician of Amiens, situated also on the Somme, who had been very sceptical about these discoveries, was induced to visit Abbeville, where he examined M. Perthes' collection; and, returning to Amiens, soon discovered similar remains in the gravel-pits near that city. He obtained, from beds containing the remains of "the mammoth, an extinct rhinoceros, and other animals now foreign to Europe, four hundred relics, most of them in silex, and wrought with singular skill. They consisted of hatchets, poniards, knives, triangular cones, besides perforated globes, seemingly beads for necklaces and bracelets." Up to this time, more than a thousand implements have been found in the Valley of the Somme in an area fifteen miles long.

Notwithstanding the evidence furnished by these discoveries, considerable doubt was expressed regarding them, especially by the geologists of England. The doubt was dispelled, however, by the visits of Messrs. Prestwich and Lyell of England, and Gaudry of France.

In the report of Mr. Prestwich to the Royal Society, he says he found the gravel-beds capping a low chalk hill, and not commanded by any higher ground. The upper bed consists of from ten to fifteen feet of brown brick earth, containing old tombs and coins, but no organic remains. Under this is a whitish marl and sand with recent shells, and mammalian bones and teeth, whose thickness varies from two to eight feet. Lastly, there are six to twelve feet of coarse subangular flint gravel, with some remains of shells in sand, and teeth and bones of the elephant, horse, ox, and deer: with these the flints were found. Mr. Prestwitch was satisfied, though he says he undertook the examination full of doubt.

"Doubt, but examine," would be an excellent motto for other fellows besides fellows of the Royal Society.

Lyell, who obtained seventy of these flint weapons, expresses himself strongly in favor of their great antiquity. Referring to the disappearance of the elephant, rhinoceros, and other genera of quadrupeds now foreign to Europe, he says it "implies a vast lapse of ages, separating the era in which the fossil implements were framed, and the invasion of Gaul by the Romans."

M. Gaudry, a member of the French Institute, accompanied by three other gentlemen, visited the quarry at St. Acheul near Amiens, and caused it to be opened for the length of about seven yards. He watched the whole operation, and never left the ground while the work went on. They found an overlying bed eleven feet in thickness, in which nothing was found; below this a flinty bed, reposing on white sand, from which nine flint implements were obtained. The edges of the flints were sharp; and in the same bed, at a little distance, were found remains of the rhinoceros, hippopotamus, and mammoth.

Mr. Lubbock, in his article on the Antiquity of Man, in "The Natural-history Review," says, "He must have seen the Somme running at a height of, in round numbers, one hundred and fifty feet above its present level. From finding the hatchets in the gravel up to a level of a hundred feet, it is probable that he dates back in Northern France almost, if not quite, as long as the rivers themselves. The face of the country must have been indeed unlike what it is now. Along the banks of the rivers ranged a savage race of hunters and fishermen; and in the forests wandered the mammoth, the two-horned, woolly rhinoceros, a species of tiger, the musk-ox, the reindeer, and the urus."

Let us transport ourselves to those primitive times, when these rude men with their stone weapons lived on the banks of the European rivers, and fished along the coast. The seasons are fairly established; and spring follows winter, and fall summer, as now; though the summer is longer and warmer than we are accustomed to see in those countries at the present time, and the winters colder. The country is covered with dense forests, through which ramble mighty elephants in herds, with immense curved tusks, coats of long, shaggy hair, and flowing manes. How proudly march these aristocrats of the elephant family! Shuffling along comes the great cave bear from his rocky den, as large as a horse: fierce, shaggy, conscious of his strength, he fears no adversary. Crouched by a bubbling spring lies the cave tiger; and, as the wild cattle come down to drink, he leaps upon the back of one, and a terrible combat ensues. It is as large as an elephant, and its horns of enormous size; and even cave tigers could not always master such cattle as they.

Are these the highest forms of life that the country contains? What being is that sitting on yon fallen tree? His long arms are in front of his body, and his hands between his knees; while his long, hairy legs are dangling down. His complexion is dark as an Indian's; his beard scanty, while his unkempt hair hangs down in snaky locks; his eyebrows overshadow his eyes; so that, with his sloping forehead and brutal countenance, he seems like the caricature of a man, rather than an actual human being.

Beneath the shade of a spreading chestnut we may behold a group, — one old man with a short, thin beard, and women and children, lounging and lying upon the ground. How dirty! What forbidding countenances! — more like furies than women. One young man with a stone axe is separating the bark from a neighboring tree, to cover the wretched hut in which they shelter themselves during a storm. Others, like monkeys, are climbing the trees, and passing from branch to branch, as they gather the wild fruit that abounds on every side. Some are catching fish in the shallows of the river, and yell with triumph as they hold their captives by the gills, dragging them to the shore. Their language of signs, and rude, undistinguishable sounds, gives little promise of the polished tongues of Europe.

Rude as this conception represents the primitive man, facts warrant the assumption that the early "stone men" of Europe were even in a more brutal condition than this. Man is in harmony with all other organized existences on the planet; and their appearance in constantly advancing forms, from the Cambrian period to the present, shows us the necessarily rude condition of the primitive man. As the earliest known fish are small and low in organization; the early reptiles, of the lowest order of reptiles; the early mammals, undeveloped marsupials: so the early men were rude, brutal, savage men, requiring ages for their development into civilized and enlightened people. We are not left, however, merely to this inference: facts bear strongest testimony to this important fact. Marcel de Serres says, "The human heads discovered in divers localities of Germany (in caves or in ancient diluvial deposits) have nothing in common with those of the present inhabitants of the country. Their conformation is remarkable in that it offers a considerable flattening of the forehead, similar to that which exists among all savages who have adopted the custom of compressing this part of the head.

"Those found near Baden, in Austria, present strong analogies with those of the African or negro race; while those from the banks of the Rhine and Danube resemble the Caribs, or the ancient inhabitants of Chili and Peru.

"In caves near Liege were found portions of the skull and other bones of man along with bones of an extinct bear, hyenas, lions, &c. The pieces of skull show that the forehead was very short, and much inclined." Professor Spring says, "The form of these crania approaches more nearly that of negroes, and not at all to present European races. The occipital bone is higher, the lateral sides of the skull much more flattened and more compressed than in any of those of our living races."

There can be no doubt that Europe was once inhabited

by savages as wild and barbarous as the Digger Indian or the Fuegian, and whose intellectual range was equally as narrow, or even more so, as their brain-cases testify.

In 1857, a human skeleton was found in a limestone cave in the Neanderthal, near Düsseldorf, Prussia. The floor was covered to the depth of four or five feet with a deposit of mud, mixed with rounded fragments of chert. It was associated with some remains of a bear, and is now believed to be of the same age as the extinct animals — such as the mammoth, rhinoceros, &c. — whose remains have been so frequently discovered in drift-deposits. Professor Huxley says "that it is, beyond doubt, traceable to a period at which the diluvium still existed. The thickness of the bones was very extraordinary, and the elevation and depression for the attachment of muscles were developed in an unusual degree. Some of the ribs also were of a singularly rounded shape and abrupt curvature, which was supposed to indicate great power in the thoracic muscles." The forehead is narrow and low (Fig. 54). When exhibited at a scientific meeting at Bonn, doubts of its human character were expressed by several naturalists. Professor Schaffhausen and other experienced geologists declared it to be human, but were struck with the resemblance of the frontal bone to that of the chimpanzee and gorilla. It is certainly the most ape-like human skull ever observed by scientific men. It is, in places, three-quarters of an inch thick; and the bony ridge over the eyes is enormous.

Fig. 54.

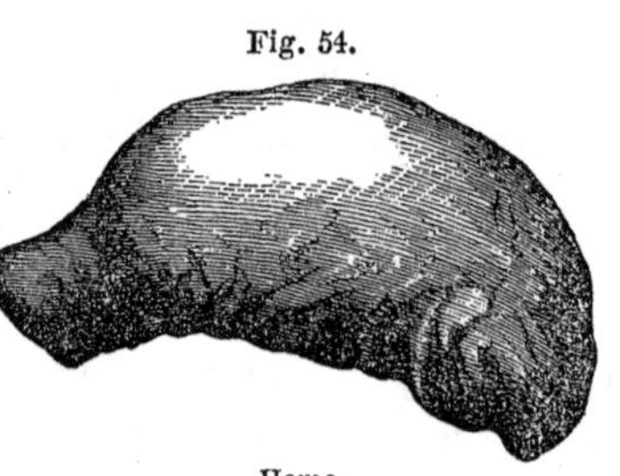

Homo.

Recent discoveries in the bone caves of Gibraltar show that this skull is not alone. The greater part of a cranium was found in bone breccia, in Cochrane's Cave. "This cranium," says Mr. Busk, "resembles in all essential particulars, including its great thickness, the far-famed Neanderthal skull. Its discovery adds immensely to the scientific value of the Neanderthal specimen, if only as showing that the latter does not represent, as many have hitherto supposed, a mere individual peculiarity, but that it may have been characteristic of a race extending from the Rhine to the Pillars of Hercules."

The aboriginal inhabitants of America were preceded by others lower in the scale than themselves. In a bone cave in Brazil, Dr. Lund discovered human crania mixed with the bones of extinct animals, in which the forehead receded on a level with the face. Castelnau discovered, under similar conditions, in the rocky caverns of the Peruvian Andes, human crania of a similar form.

North America seems to have been peopled also by rude races at a very early period, probably preceding the ancient dwellers on the banks of the Somme, whom Lyell estimates to have lived one hundred thousand years ago. A human skull was found by Mr. James Matson, at Altaville, in Calaveras County, California, at a depth of one hundred and thirty feet, under four beds of consolidated volcanic ash, locally known as lava, associated with the bones of an extinct rhinoceros, camel, and horse. It was covered and partly incrusted with stony matter. The base of the skull is embedded in a mass of bone breccia and small pebbles of volcanic rock, cemented with a thin layer of carbonate of lime. The skull resembles those of the present Digger Indians, but is of remarkable thickness; a feature indicating small use of the brain on the part of its original owner, mental exercise reducing the thickness of the skull.

It must not be supposed that we have yet found remains of the earliest human beings. It is highly probable that man existed during the pliocene period, when we know that the horse, the ox, the deer, the dog, and other familiar animals, abounded.

The immense weight of the ice accumulated during the drift-period seems to have depressed a large portion of the land below the ocean-level, owing, probably, to the still yielding nature of the earth's crust; and in some places the ocean continued to occupy the land for some time after the icy covering melted. Evidences of this are common along the coast of New England, around the shores of Lake Champlain, and in various parts of Canada, near the St. Lawrence and its tributaries, and in Nova Scotia. At the close of the drift-period, the land slowly rose step by step, leaving terraces, marking the former beaches along which the waves rolled, as I have seen them on the Lower St. Lawrence, one above the other, to a height of three hundred feet; attesting the successive uplifts to which the land has been subjected. Around Lake Champlain there has been an elevation, in some places, of more than three hundred feet, and above four hundred near Montreal; for beds are found at these elevations abounding in marine fossils.

In making the excavation for the Rutland and Burlington Railroad, a fossil whale, about fourteen feet long, was found in blue clay, fourteen feet below the surface, sixty feet above the present level of the lake, and one hundred and fifty feet above the ocean. Seals, fishes, and immense num' ers of shells, have been discovered in beds around

Lake Champlain, deposited after the glacial period, and before the uprising of the land took place; the ocean passing up the St. Lawrence (then a long gulf), and connecting with Lake Champlain, a large inland sea.

That this submergence of the land was caused by the weight of the ice is indicated by the fact that the greatest amount of submergence took place in the arctic regions, where the icy covering was necessarily the thickest, for there the old sea-beaches are highest: they are higher on the St. Lawrence than in Southern New England; and the northern terraces on the Great Lakes are higher than those farther south.

ALLUVIAL PERIOD.

More recent, generally, than the drift-beds, are the alluvial deposits, some of which are in process of formation to-day. The world is by no means finished yet. Ever the water flows, the wind blows, the grass grows, the earthquakes heave, the waves dash; ever the destroyers pull down the old world, ever the builders erect the new. Among the sayings attributed to Pythagoras by Ovid, we have the following: "Nothing perishes in this world; but things merely vary, and change their form." As evidences of this, he says, "Solid land has been converted into sea. Sea has been changed into land. Marine shells lie far distant from the deep, and the anchor has been found on the summit of hills. Valleys have been excavated by running water, and floods have washed down hills into the sea. Marshes have become dry ground. Dry lands have been changed into stagnant pools. Volcanic vents shift their position: there was a time when Ætna was not a burning mountain, and the time will come when it will cease to burn." Aristotle says, "The distribution of land and sea in particular regions does not endure throughout all time; but it becomes sea in those parts where it was land, and again it becomes land where it was sea. The same tracts, therefore, are not, some always sea, and others always continents; but every thing changes in course of time."

Mohammed Kazwina, an Arabian writer of the seventh century, embodied the idea in a fine conception, which has been thus translated and versified:—

"I wandered by a goodly town
Beset with many a garden fair,
And asked, of one who gathered down
Large fruit, how long the town was there.
He spoke, nor chose his hand to stay:
'The town has stood for many a day,
And will be here forever and aye.'

A thousand years went by, and then
I visited the place again:
No vestige of that town I traced.
But one poor swain his horn employed:
His sheep, unconscious, browsed and grazed.
I asked, Whèn was the town destroyed?
He spoke, nor would his horn lay by:
'One thing may grow, and another die:
But I know nothing of towns; not I.'

A thousand years went by, and then
I wandered past the spot again.
There, in the deep of waters, cast
His nets a lonely fisherman;
And, as he drew them up at last,
I asked him how the lake began.
He looked at me, and laughed to say,
'The waters spring forever and aye,
And fish are plenty every day.'

A thousand years went by, and then
I saw the selfsame place again.
And, lo! a country wild and rude;
And, axe in hand, beside a tree,
The hermit of that solitude,
I asked how old the wood might be.
He said, 'I count not time at all;
A tree may rise, a tree may fall;
The forest overlives us all.'

A thousand years went by, and then
I went the same old round again;
And there a glorious city stood;
And, 'mid tumultuous market-cry,
I asked, When rose the town where wood,
Pasture, and lake forgotten, lie?
They heard me not; and little blame.
For them the world is as it came;
And all things must be still the same.

A thousand years shall pass, and then
I mean to try that road again."

And, if he does, the sea may be there, and gallant ships go sailing by. So goes the world, though multitudes have no conception of it. At the close of one of my lectures, a gentleman came up, and remarked to me that he believed the world was made "just exactly as it is." — "What!" said I, "down to the cart-ruts?" — "No-o: not exactly down to the cart-ruts." — "Then not to the river-ruts; for, as the carts have ruts to run in that their wheels have made, so the rivers have ruts to run in that their wheels have made." See what Niagara has done! Six hundred and seventy thousand tons of water falling every minute over a limestone ledge is wearing it back at the rate, says Lyell, of a foot a year. Observers on the spot, whom I have conversed with, say less. But suppose it a foot a year: it has then taken thirty-five thousand years to cut its way back from Queenstown, seven miles below, where it once was, to Niagara, where it is now.

Below Rochester, the Genesee has cut its way back for seven miles through rocks of three different degrees of hardness, that have worn away with unequal rapidity; so that there are three cataracts, miles apart. These are trifling operations compared with what water has accomplished in other places. In New Mexico, Texas, Colorado, and California, there are long, deep gulfs, called "canyons," where, for days together, the traveller can find no place to cross; and, though water is in sight, there is no chance to obtain it. Grand Canyon, on Canadian River, is two hundred and fifty feet deep, and fifty miles long. Capt. Marcy describes a canyon in Texas, on Red River, seventy miles long, and from five hundred to eight hundred feet deep. In some places on the Colorado, they are more than a mile deep, — a thousand feet in solid granite.

But what is this compared with the disintegrating action of the ocean, incessantly battering its banks, and making continual inroads on the land? At Cape May, it was proved by measurement, that, from 1804 to 1820, the sea encroached on the land at the rate of nine feet a year. At Sullivan's Island, near Charleston, in South Carolina, the sea carried away a quarter of a mile in three years. Near Boston Harbor, the islands are rapidly wasting away under the influence of the Atlantic waves; and on one of them, Deer Island, an extensive wall of stone has been erected to arrest the work of destruction. It has been calculated that one thousand tons of earthy matter are transported daily from the coast of Long Island, seaward.

The des ructive agency of the sea can be most readily seen in Nova Scotia, on the Bay of Fundy. From Briar Island, in the south, to Cape Blomidon, in the north, the amygdaloidal traps, which form in many places cliffs several hundred feet high, are rapidly wasting away. Every spring, immense masses of rock are dislodged from them, and the *débris* reduced by the waves to mud, and swept seaward. The trap displayed at Cape D'Or and Partridge Island, on the western side of the Bay of Fundy, was once continuous with that on the eastern side. Cut through by a river, probably, at first, it has since been wasted away by the action of the waters of the Bay, till at Cape D'Or there is a width of seven miles; and a few centuries will destroy all vestiges of volcanic agency on the western side of the Bay of Fundy.

Crossing the ocean to the Island of Great Britain, "where they measure land by the inch," we can obtain a better idea of what the ocean is doing, because greater attention is paid to its ravages. Land is dearer, population more dense; and intelligent persons have been watching the process, and recording it, for hundreds of years. North of Scotland, the Orkney and Shetland Islands, though made of the hardest and most enduring materials, have been worn away to mere needles. On the coast of Durham, between Sunderland and Hartlepool, where magnesian limestone cliffs of the Permian formation front the German Ocean, there are caves innumerable, made by the rolling waves. In a stormy time, the waves lift up masses of rock, that weigh hundreds, or even thousands, of pounds, and play with them as a boy does with a ball. Against the face of the cliff they are driven with terrible force: if one place is softer than another, it is soon discovered, and a cave is in time hollowed; where two exist contiguous, the space is eaten away between them, the unsupported roof falls, and the sea has made a permanent inroad into the land.

Proceeding southward along the eastern coast of England, we come to Yorkshire, the cliffs of which are crumbling away along its whole extent. At Whitby there is an old abbey, or was when I was there, perched on the edge of the cliff. Ask why they built it so close to the cliff, and they will tell you, that, when it was built, it was a mile and a half inland, and the sea has swallowed the land that lay between them. Auburn, Hartburn, and Hyde are towns known only in history and tradition: their bones lie beneath the German Ocean; and fishes swim where men have been. At Owthorne, the annual encroachment for many years was about twelve feet. The cliffs of Norfolk and Suffolk are wearing away with great rapidity. The ancient villages of Shipden, Wimpwell, and Eccles, have disappeared. The whole site of the ancient town of Cromer has been swept away, and ships sail over where the houses were. South of this, the cliffs, it is estimated, are going at the rate of three feet annually.

Dunwich, in Suffolk, was once a place of much importance, containing at one time fifty-two churches and monasteries. A monastery went at one time, at another several churches, and subsequently four hundred houses at once. "Ancient writings make mention of a wood a mile and a half to the east of Dunwich, the site of which at present must be so far within the sea." The sea is now twenty-four feet deep where the town of Aldborough once stood.

Farther south, we come to the mouth of the Thames and the Isle of Sheppey, which is composed of London clay, and decays very rapidly. In 1780, the Church of Minster, now near the coast, is said to have been in the middle of the island. In a few years, some say thirty, the island, which is about six miles long by four broad, will be taken into the never-to-be-satisfied maw of the sea. In the county of Kent, we find Reculver, with its ancient church, on the very edge of the cliff: in the time of Henry VIII., it was a mile distant. In 1834, wooden piles were driven, and a stone breakwater built, to save the church. The sea never relaxes its vigilance for a moment, and will some day seize its prey. In nearly every instance where cliffs abut on the sea, they are being worn away by the action of the waves, however hard the material may be of which they are composed. There is little doubt that England and France were once united, probably within the human period; and the division that an earthquake may have commenced, the sea has enlarged, and is constantly increasing.

Some parts of Holland have been terribly devastated by the sea. North Friesland was, in the year 1240, a flourishing district ten miles long, and seven miles broad. Toward the end of the sixteenth century, it was only four miles round. In the year 1634, a flood passed over the whole island, when thirteen hundred houses with many churches were lost: fifty thousand head of cattle perished, and above six thousand men. Three small islets remained; but they are wasting away.

Is this, then, to be the fate of the world? Must the hungry sea swallow all, and the waves chant forever the funeral dirge of the lost, lost land? There is no danger. The universe abounds with checks and counterbalances:

the existence of the earth as it is, and man upon it, is evidence of this. More rapidly than the sea eats away the land, the land gains upon the sea. There is a greater land-surface on the face of the world to-day than at any past period in its history, and this is constantly increasing. How is this done? Away up in the mountains, the rain and the frost loosen a massive rock from its seat of ages, and down it falls into the bed of a torrent. In time, it is broken into pieces as large as a flour-barrel; and these are swept down, rolling one against the other, till they are no larger than a man's fist. Now they are swept along with the greatest ease; for a stream flowing at the rate of four miles an hour can do that: on they go, till they are ground to sand, and then to impalpable powder; never resting till they are swept into the ocean, and settle down to make new rock, some day to become dry land, and be inhabited.

> "Lofty mountains, whose tops appear to shroud
> Their granite peaks deep in the vapory cloud,
> Worn by the tempest, wasted by the rains,
> Sink slowly down to fill wide Ocean's plains."

Every shower washes material down from the hills into the plains, and thence eventually to the ocean; and, where the earth is loose, the streams become so charged with it, as I have seen them in Western Colorado, as to flow with mud instead of water.

The Mississippi displays on the grandest scale the action of running water. Three thousand miles from its source, "where the fur-hunter roams the wilderness for his peltries," it pours its turbid waters into the Gulf of Mexico in the region of the cotton, rice, and sugar. Its alluvial plain is from fifty to ninety miles wide, made by the river swinging from side to side like a mighty pendulum, marking the seconds of the grand geologic periods. The land formed by the sediment carried down in its waters contains thirteen thousand six hundred square miles. It is said to gain on the Gulf at the rate of one mile in a century. The outer crest of the bar of the South-west Pass advances into the Gulf three hundred and thirty-eight feet annually. The coarser sediment is deposited near the mouth: but the finer is swept on by the current of the stream, slowly sinking as it goes; and is distributed widely over the sea-bottom, there to be consolidated in time into rock, when the pressure of the sediment above it becomes sufficiently great, and time has been given to harden it. It has been estimated that four billion cubic feet of mud are carried down every year, or sufficient to cover a township, containing thirty-six square miles, to the depth of forty feet. It would take forty thousand locomotives, taking a thousand tons a trip, and making a trip a week, to transport as much solid material as this mighty river sweeps down suspended in its waters.

The Amazon of South America probably carries down a larger amount of sediment than the Mississippi; but the Gulf Stream, sweeping past its mouth at the rate of four miles an hour, prevents the formation of a delta, and carries its sediment out into the ocean. Its turbid waters can be distinguished three hundred miles from its mouth; and the fine sediment must be transported much farther than this.

It is generally where rivers empty into a lake, sea, or ocean, that deposits of land are now being made. In most cases, they are called "deltas," from the Greek name for the letter *d*, to the shape of which (Δ) they bear some resemblance.

That Egypt was the gift of the Nile was a common expression in that country more than two thousand years ago. It is supposed that the sea once washed the base of the rocks on which the Pyramids of Memphis stand; and all the land lying between has been deposited by the river. It has swept down from the mountains of Abyssinia the fine mud that lines its fertile valley, and has formed a delta at its mouth as large as the State of Vermont.

The Lake of Geneva is about forty miles long, and from two to eight broad. The waters of the River Rhone enter it at the eastern end, turbid, and highly discolored, but leave it at the western end, near Geneva, beautifully clear and transparent. Thus the lake is filling up with the sediment deposited from the waters of the river: so that an ancient town called Port Vallais, once situated at the water's edge, near the eastern end, is now a mile and a half inland; that amount of land having been made in about eight centuries. In time, the whole of the lake will be filled, and a rich agricultural district take its place. Millions of acres that farmers now plough have been formed in a similar manner.

The Ganges and Brahmapootra, the two principal rivers of India, descending from the highest mountains in the world, and passing through a country where the rain-fall is very great, take down an immense amount of solid material. There is nothing as fine as gravel within four hundred miles of the mouth of the Ganges; it is all fine mud brought down by the stream: while the united delta of these streams is as large as the State of New York. It has been computed that the Ganges alone carries into the Bay of Bengal, every year, seven thousand million tons of mud, or, in other words, a mass of material more than sixty times as large and heavy as the Great Pyramid of Egypt, whose base covers eleven acres, and whose height is four hundred and fifty feet.

The Po and the Adige, two European streams flowing into the Adriatic Sea, or Gulf of Venice, have caused a hundred miles of coast to encroach on the gulf from two to twenty miles in two thousand years. The city of Adria, which was a seaport in the time of Augustus, and gave its name to the gulf, is now twenty miles inland. Thus the Gulf of Mexico, the Bay of Bengal, and the Gulf of Venice, are steadily filling; being slowly converted into dry land, which will represent this period, as the old formations represent the periods of the past. During the war, down sank into the bed of the Mississippi the boats of contending parties; side by side lie the bodies of Federal and Rebel; coins, jewels, clocks, watches, pottery, cutlery, books on all subjects, and a thousand machines and implements. The fine mud settles around them as they sink; and in time many will become fossils in the solid rock. When our libraries shall have perished, our newspapers be forgotten, and our most cherished names be lost in the millions succeeding them, even then Nature shall preserve her faithful diary; for there is nothing too trivial for her notice, too slight for her regard. At some time, the bottom of the Gulf of Mexico will be elevated above the sea-level; rivers will course over it, exposing the rocks making to-day on the right hand and on the left; men will open quarries in them to build cities, and the works and remains of this age will come to light. I can imagine with what feelings of astonishment the scientific men of that age will view a fossil hoop-skirt or a capacious Dutch tobacco-pipe, and what singular notions they will entertain of the semi-barbarians who live in these days.

In the play of "Richard III.," George, Duke of Clarence, relates to Brakenbury, in the Tower, a fearful dream: —

"Methought I saw a thousand fearful wrecks;
A thousand men that fishes gnawed upon;
Wedges of gold, great anchors, heaps of pearl,
Inestimable stones, unvalued jewels, —
All scattered in the bottom of the sea.
Some lay in dead men's skulls; and, in those holes
Where eyes did once inhabit, there were crept
(As 'twere in scorn of eyes) reflecting gems,
That wooed the slimy bottom of the deep,
And mocked the dead bones that lay scattered by."

In the Museum of the Boston Natural-history Society are fossil wine-decanters and a ship-bell. I have seen English guineas in conglomerate, taken from the wreck of the British man-of-war "Huzzar;" the solid conglomerate, formed by oxide of iron, binding together the flint pebbles of which the ballast was composed. I have seen silver pennies of Edward I., in breccia, found at the depth of ten feet in the bed of the River Dove, in Derbyshire, in 1832. They came from a military chest lost in this stream in 1322: the soldiers, being alarmed by a sudden panic, threw the chest into the river. Many thousand silver coins, English, Irish, and Scotch, were found in a hard conglomerate. The full significance of the fossils formed, and thus forming, we are far from comprehending yet: they will convey to the future student much more than the geologist at present dreams of.

There is no danger, then, that the sea will swallow up the land: the land, in fact, gains on the one hand more than the sea sweeps off on the other. But will not the degrading influences, whose existence and operation have been demonstrated, reduce the earth to a dull, monotonous plain? The rain, falling on the mountains, is constantly sweeping fine particles into the valley, and there are no streams to return them: frost and ice are equally employed in wearing down the high places. Millions of laborers, day and night, employed in levelling the earth, what can withstand their power? They have been at work for millions of years, and yet the hills and mountains survive; or, if thousands have been levelled, thousands more have taken their place. A cooling globe, such as the earth is known to be, is necessarily a shrinking globe. A shrivelled apple is such in consequence of the inside becoming too small for the skin on the outside, which shrivels to accommodate itself to the changed conditions. The crust of the earth was first formed when the earth was much larger: it fitted the earth at first, as the skin does a green apple; but, as the earth diminished in consequence of parting with its heat, its rocky covering has been ridged into hills and mountains, and shrunk into valleys. Thus the crust of the earth is constantly rising or falling, accommodating itself to the shrinking interior, the heat of which is passing off by volcanoes, hot-springs, and by conduction through the rocks to the surface. Thus it is known that the western coast of Greenland, for a space of more than six hundred miles, is slowly sinking. "Ancient buildings on low, rocky islands, and on the shore of the mainland, have been gradually submerged; and experience has taught the aboriginal Greenlander never to build his hut near the water's edge. In one place, the Moravian settlers have been obliged more than once to move inland the poles upon which their large boats were set; and the old poles still remain beneath the water as silent witnesses of the change."

For many years, a discussion was carried on regarding the alteration of level on the peninsula of Sweden and Norway. To decide the matter, lines, or grooves, were cut in the rocks at the water's level on the shore of the Baltic. If the land should rise or sink, since there are no observable tides on the Baltic, those marks would tell the story. Fourteen years afterward, they were examined by Lyell, who found there had been a rise of four or five inches. This rise seems to have been going on for a long period, as masses of shells have been found on old sea-beaches two hundred feet above the present level of the sea, containing only such species as now live there. South of Stockholm, the peninsula is saidto be slowly sinking.

At Spitzbergen, drift-wood, and bones of whales, have been found several miles inland, and at least thirty feet above high-water mark.

"The coast of Denmark on the Sound, the Island of Saltholm, opposite to Copenhagen, and that of Bornholm, are rising, — the latter at the rate of a foot a century. The coast of Memel, on the Baltic, has actually risen a foot and four inches within the last thirty years."

In England and Scotland, the land has in many places been rising for ages, as the presence of elevated sea-beaches along the coast testifies.

No better evidence can be found of the elevations and subsidences to which the earth's crust is subject than that of the Temple of Jupiter Serapis at Puzzuoli.

Originally, the temple had twenty-four granite columns, and twenty-two of marble, each hewn from a single stone. Three of these columns remain erect, the tallest forty-two feet high. They are smooth and uninjured to the height of twelve feet; but above that are bands of holes nine feet wide, made by the boring of sea-mussels, some of the shells still occupying them. Above these bands, the columns are smooth to their summits. It is evident that the pillars were once below the level of the sea, their lower portions being protected by rubbish; and since that time they have been elevated twenty-three feet, without disturbing the columns, which, as Moore says, —

> "Stand sublime,
> Flinging their shadows from on high,
> Like dials which the wizard Time
> Had raised to count his ages by."

There is reason to believe that Newfoundland is rising out of the sea. In the neighborhood of Conception Bay, large, flat rocks exist, over which schooners used to pass thirty or forty years ago: now the water over them is scarcely navigable for a skiff.

The southern coast of Nova Scotia is rising. Near Salmon River are bluffs fifty feet high, containing shells, most of which are identical with those in the contiguous sea.

At Tadousac, on the St. Lawrence, below Quebec, are old sea-beaches from fifty to three hundred feet above the present level of the water. They have been favorite places of resort for old Indian tribes: their stone weapons lie scattered over them in the greatest abundance.

Near Lima, in South America, Darwin found proofs that the ancient bed of the sea had been raised more than eighty feet within the human period; strata having been found at that height above the sea-level, containing pieces of cotton thread and plaited rush with seaweed and marine shells.

Sometimes elevations and subsidences are made by earthquake agency; and at such times the land rises or sinks with great rapidity. In 1819, an earthquake occurred at Cutch, in the delta of the Indus. The fort and village of Sindree were submerged, the tops only of the houses appearing above the water. Two thousand square miles of land were converted into a lake; but, immediately after the shock, a mound fifty miles long, fifteen miles wide, and ten feet high, arose within a few miles of the place of subsidence.

In 1822, the coast of Chili, after an earthquake, was permanently elevated from two to six feet, through an area calculated to be one hundred thousand square miles in extent.

We shall have, then, for the ages to come, the giant mountains with their snowy crests, lifting up the souls of millions yet to be born, and giving them visions of beauty and grandeur. We shall have our verdant hills, from which shall go laughing rills on their beneficent mission; and flowery vales, where the lovers of the coming time shall walk and rejoice in the varied prospect spread before them.

The continual changes taking place in the inorganic world are accompanied by, and to some extent productive of, continued change in the organic world. The mastodon, the mammoth, the cave bear, the Irish elk, the woolly-haired rhinoceros, and many other animals, have become extinct within the human period; some of them, such as the dodo and solitaire, within a few centuries. Many vegetable forms have doubtless perished in like manner; though, being less conspicuous, and more difficult of preservation as fossils, we are not as well acquainted with those that have become extinct.

Accompanying this extinction of the old is the creation of the new. New varieties, which it is now generally acknowledged are incipient species, come into being every year. Many perish; some live, struggle, perpetuate, and eventually give us new species of animals and plants better adapted to present conditions than the old. Thus is the face of the earth renewed from age to age, and its continual progress guaranteed.

LECTURE VI.

THE FUTURE OF THIS PLANET AND ITS INHABITANTS.

WIDE as humanity, and inseparable from our very nature, is the disposition to look into the future. As the traveller on the highway eagerly views the road that lies before him, so the travellers on life's pathway climb the hills, and peer through the clouds and fogs, to catch a glimpse of the path that they are destined to walk. To gratify this disposition, the astrologer consults the stars in nightly watches, and interrogates them regarding the influence he supposes them to have on human destinies; the cheiromancer scrutinizes the lines of the hand to find some connection between them and the lines of individual fate; the necromancer seeks to call up the very dead, and wrest from their skeleton fingers the key that shall unlock the secrets of the future.

Nor is this disposition confined to the vulgar. It is shared in alike by the peasant and the prince, the poet, the prophet, and the philosopher; nor without avail. In every age there have been men and women, who, soaring into the sky like larks, have seen the rays of the rising day before its beams have reached the earth, and sung of its glories in the ears of the drowsy world lying in the twilight below.

The World holds her hand to us; and we read the lines that the ages have carved thereon, and from them unhesitatingly state what will be her coming history. She gives us, in geology, all the necessary elements; and we calculate her nativity.

Can any one really tell what will be? Can any unassisted mortal reveal what the future has in store for our planet? Assuredly he can, with such assistance only as Nature gives to her faithful students. As certainly as we can tell, when looking on the gigantic forest-tree, the pride of the woods, that the time will come when it shall lie prostrate on the ground, and give back to the earth and the air all that it gathered from them; as certainly as we can tell that the lordly city, whose merchants are princes, whose dwellings are palaces, whose ships are on every sea, shall die, and the place where it sits in its majesty shall be a country wild and rude (for cities have their time to fall as truly as trees); as certainly as we can tell when an eclipse shall take place, and, at the very moment designated years beforehand, the shadow commences to creep over the face of the sun or moon: so certainly can we tell, in many respects, what is the destiny of this globe and of man dwelling upon it.

The past is the certain guide to the future. He who is best acquainted with the past is the best prepared to tell the future; and that science which most clearly reveals to us what has been is the science to make known to us what is yet to be. That science is geology. It casts a light upon the future radiant as the sun, dispelling the clouds that curtain the distant land, and enabling us to mark its mountain-chains, trace the course of its principal streams, and observe its most prominent features.

We stand on the mount of a myriad years,
And view with a prophet's ken
The course of this onrushing world to its goal,
The fate of its future men.

The first statement that geology enables us to make in reference to the future of this planet is, that it will continue for millions of years, as it has been in existence for millions. While men believed that the earth had attained its present maturity in less than six thousand years, it was not unreasonable to suppose that it might very shortly perish. If so stately a tree had grown to its prime in so short a time, a few years more might lead it down to the grave; but, when we learn the story of its mighty past, our conclusion is very different indeed.

The first vegetable forms with which geology makes us acquainted are naked seaweeds, destitute of leaves and branches; and many millions of years passed before exogenous trees adorned the earth, or fruit-trees bore their blushing load. The earliest remains of fishes, as yet discovered, are found in the upper Silurian. They are small in size, and inferior in organization; and it is not until we arrive at the carboniferous period, certainly millions of years after their first appearance, that we find fish developed to the highest type of which fish-life is capable. The earliest undoubted remains of reptiles so nearly resembled fishes in their structure, that Agassiz was deceived with regard to their true character; regarding them as fishes instead of salamandroid reptiles, or reptiles resembling the salamander. The skeleton was somewhat cartilaginous, or gristly, while they belong to the batrachia, the lowest order of reptiles. Millions of years elapse before the gigantic ichthyosauri float upon the waters, the iguanodons march through the forest, the crocodiles crawl along the shores of the rivers, and reptiles attain their most perfect forms. The earliest mammals are from the triassic beds of Germany, — small insect-eating marsupials, belonging to the lowest order of their class, and the most reptilian in character; and millions of years roll along through the oölitic, cretaceous, and tertiary periods, before the horse scours the prairie, the elephant roams through the jungle, the cow browses in the pasture, and the monkey chatters in the tree.

If Nature gave the fish such an immense period of time in which to unfold its cold-blooded, pygmy progeny to their full stature, millions of years to perfect crawling monsters, and millions more to advance beasts from the small-pouched mammals of the trias to the mastodons and monkeys of the tertiary, will she not give time to humanity, her master-piece, to arrive at perfect manhood? Of course, and time to enjoy it after this consummation, so much to be desired, has been attained. What a vast period this must necessarily be! A perfect man the world has yet to see. Far apart, with centuries of barrenness between, noble men appear, and show us the possibilities of the race; but how few they are! We might count them on our fingers, and then have digits to spare, — so few, indeed, that we are ready to bow down and worship them as gods. What ignorance among the scientific, and yet what pride! — as if the universe had no new pages to be read. What uncharitableness, selfishness, and even inhumanity, among the professedly religious, whose love is too often the offspring of their fear! Among politicians, how few that are honest and true! Their cry is, "The people, the people!" but their aim, power and pelf. What draws the crowd in a city? A monkey, it may be; or a troop of men who have made themselves as like monkeys as possible by a lifetime of hard training. What are the objects of pride among civilized men? A title, a ribbon, a cross, a handsome pipe, or a fast horse; among women, a "love of a bonnet," a splendid mansion, a wealthy husband, or entrance into fashionable society. How few make the development of the mind the first business of life! We are a race of babies, it must be confessed: doctors, lawyers, parsons, lecturers, princes, presidents, — babies all! How long will it be before these millions, toiling, moiling, like ants on a hill, or caterpillars on a branch, shall become men? If a hundred thousand years at least have been spent in bringing us to where we are, how many must be given to carry us as high as our highest ideal of human excellence? Ages; and the earth shall have them. Nature is never niggardly of time. "Do you need a million years?" she says. "Call on me, and fear not. I gave them to the fish, who never desired them; to the reptile, who cared not for them; to the beast, who heeded them not: how much more shall I give them to you! You cannot overdraw at my bank; for eternity is mine, and all of it that is needed is yours."

Strange, there are men who dream that the course of this planet is nearly run, though it is yet so far from the goal! — not strange either, when we think how we have neglected Nature's great volume of instruction, and listened for ages to fables. There has not been a year for the last eighteen hundred that many have not looked forward to as the last of the expiring world. Not a meteor's glare, nor an earthquake tremor, but is hailed as a herald of coming chaos: and yet the grand old earth spins round and round, carrying these people along with it to their destiny; and so it will do for all their brethren yet to be born. Ask them what they think the world was made for, and they reply, of course, for the production and development of men; yet, just as it commences to answer this end, they anticipate its destruction. A gentleman selects this town as the place where to build a factory for making locomotives; digs deep; lays solid foundations, and rears a suitable superstructure. Within it he places an engine, and shafts through various rooms connected with it: on the shafts are drums and belts connecting them with various machines, — some for turning, some for boring, and others for planing. After spending years of time, hundreds of thousands of dollars, and much labor, at length it produces tolerable locomotives. It takes time for the wheels to run smoothly, time for the workmen to execute their parts with accuracy. But, just as this is in a fair way to be accomplished, its proprietor burns it to the ground. What should we think of him? We should charitably conclude that he was deranged.

The earth's foundations were laid deep and enduring in the eternity of the past; and, after unceasing preparations for untold ages, the grand factory for making men out of granite commences to produce tolerable specimens of the race, with the promise of vastly better in the future: but, just as it does so, these people believe it will be burnt up, swept with universal destruction, that it may be refitted for a handful of "saints," certainly no better than the average of their neighbors, who are to occupy it forever. No danger! That the earth will cease to exist as it began to be, there is no doubt; but its end lies far away in the ages to come, when its fruit is ripe and its work is done.

A tree that takes twenty years to arrive at maturity will last for a hundred at least; and since the earth has grown during many millions of years, as we now know, we may safely calculate on its continuance for millions of years to come. Our eyes have not yet beheld the whole of its surface. Shall our inheritance be taken from us before we have seen it? We have not used the stores laid up for us in the world's cellar; nay, we are finding new ones almost every day, and therefore have good reason to believe that we have not yet discovered all the treasures prepared for us. Shall the earth be destroyed before we have received its gifts or appropriated its blessings?

The world is a noble vessel, freighted with a thousand million souls, furnished with boundless stores in her deep hold, fairly started for a distant port, every sail at last set, having the best of captains, who will hardly run her upon a rock for the sake of making a raft out of the wreck for a handful of noisy passengers, leaving the rest to perish.

After we have decided that the world shall endure for ages, the question next arises, What will be its future condition? Is it the forest monarch, its trunk rounded to its full capacity, its branches matured, its fruit perfect, the years of the future adding nothing to its glory? or is it a tree with its heart unknit, its best branches undeveloped, its beauty unmatured, its fruit imperfect, waiting for that which time alone can bring?

Old as geology represents the world to be, it still more clearly shows its youth; and the philosopher calmly waits for its improvement, as an intelligent parent does for that of his child.

> The world is young, my brothers:
> We are all here in good time.

No man ever saw the earth in a better condition for man's occupancy than it is to-day.

> "Abraham saw no fairer stars
> Than those that burn for thee and me."

Amid the countless mutations of the earth, it will still march on to its great and glorious destiny. Behind the eastern hills lie brighter days than have yet dawned, and the earth shall rejoice in their glory.

Progress is the law of our globe, as geology abundantly testifies. If we could but glance at its history for fifty or a hundred years, we might doubt it; but sweeping over the ages of the mighty past, and contrasting its early appearances with those widely succeeding, we can doubt no longer. We see it a puling infant in its fiery cradle, curtained with sulphurous clouds; then with the bare, flinty rock for its floor, and life as impossible as in a fiery furnace, its air more poisonous than the breath of a volcano, and its rain as corrosive as sulphuric acid. In time, rocks are ground to mud, and the simplest of plants spread their rootlets through it in search of nutriment. The air loses its sulphur and its carbon, which are stored away for distant uses. The water becomes purer, and all elements better fitted for the development and sustentation of life, which advances from the seaweed to the cedar, the wheat, and the rose-bush; from the unsensitive radiate, through mollusk, fish, reptile, bird, and mammal, to intelligent man. If the world has thus improved in the past, what more reasonable than that it shall continue to improve in the future? If it has marched with such an unfaltering step in the pathway of progress for such an immense period, who can doubt that it will continue so to do? Why should progress cease at this period in the world's history? If there was any reason for improvement when there was nothing to behold it but the leaden eye of the fish, or to care for it but the dull reptile, how much more reason now that man is here, eagerly watching every advance, his happiness increasing at every step of its progress!

Not only does the knowledge of the past that geology gives enable us to predict the general improvement of the earth as man's abode, but by it we can indicate more particularly the direction that this improvement will take. First, volcanoes will die, and earthquakes cease. The outpourings of fiery matter from volcanoes, and the convulsive shakings of the ground, are among man's greatest troublers: they are constant sources of apprehension to those who live in neighborhoods subject to them; and those who have been most familiar with them regard them with the greatest horror. No country is exempt from their influence, and their consequences are sometimes most disastrous. In the year 526, as many as two hundred and fifty thousand persons perished at Antioch; and, seventy-six years afterward, a second earthquake destroyed sixty thousand. In 1692, seventy-four thousand persons lost their lives at Messina; and in Quito, in 1797, forty thousand. It is estimated that thirteen millions of the human race have thus perished On an average, there is an earthquake every eight months that is destructive to human life; and of smaller and harmless ones, a shock somewhere every day. Yet how solid the earth is to-day, compared with what it once was! The rents, crevices, and veins which seam the rocks in millions of places, their sides often made smooth as glass by rubbing against each other as the rocks have been elevated and depressed, tell a story of the past world, in contrast with which the present seems peaceful and stable.

There was a time, without doubt, when the earth was a volcanic globe, its boiling lava forming a shoreless sea of fire. Long after this, volcanoes must have covered its surface as blisters do the slag of an iron furnace; its face pitted and scarred as the moon's is to-day, or a man's marked by the small-pox; while the thin crust was in continual motion by the internal, heaving tides. Ages passed, and the rocks thickened, the numerous craters were obliterated or buried, earthquakes less frequently rent it, and comparative peace and order reigned. New England, now so quiet, was once shaken and convulsed, seamed and rent: enormous crevices poured out glowing rivers, that licked up lakes and inland seas with fiery tongue, leaving marks that tell to the geologist of these desolations of an early time. The volcanoes and earthquakes of the present are but the puny offspring of a once mighty host; and are destined, like them, to expire. While active volcanoes are numbered by tens, extinct ones are numbered by thousands. In New Zealand, there are sixty extinct volcanoes within a radius of ten miles; thousands in Italy and Central France; some of them much larger than any active ones.

Day by day, the world is cooling; radiating its heat into space through its thick crust; sending it out through volcanic vents and hot-springs. Its rocky ribs increase in thickness and strength continually; and the time must come, however distant it may be, when the last earthquake shall give its last heave, and lie down in its rocky den, and expire, — when the last volcano shall send out its last smoky puff. No more shall the dwellers near Vesuvius and Ætna look up with terror to the mountain, or shake with dread at the midnight hour as they listen to the subterranean thunder. Men will plant vine-yards and olive-yards to their summits; descend into their craters, and make orchards; and young men and maidens shall dance in the moonlight, in the socket of the blind volcano's eye.

The past history of the globe furnishes indications that useless and troublesome plants will disappear, and seeds and fruits previously unknown shall bless the future dwellers on the earth. During the carboniferous period, the land-surface of the earth was a wilderness of weeds, rushes, ferns, horse-tails, club-mosses, everywhere; but no grass, no grain, and probably no edible fruit: the produce of a million acres would not have furnished a man with a breakfast. Where are the giant weeds of this olden time and of times long subsequent? All vanished. As conditions improved, vegetation improved; higher forms appeared; inferior forms died out: and the means by which this was accomplished exist and operate to-day, — our new and improved varieties of plants are evidences of it, — and will improve the plants of the future. Noxious weeds shall thus disappear, and fruits that neither we nor our fathers have known shall take their place.

The world is now infested with lions, tigers, leopards, hyenas, bears, crocodiles, boa-constrictors, vipers, rattlesnakes, and other poisonous reptiles, destroying human beings and their domestic animals. Whole villages in India have been depopulated by the ravages of the tiger; lions are the dread of man in many and large districts of country; and we are acquainted with nearly sixty species of poisonous snakes. In consequence of these, the sum total of human enjoyment is much decreased. There is great room for improvement in this direction; and it will come. Had a man been dropped upon our planet during the oölitic period, he would but have served as a mouthful for some of the huge carnivorous crawlers with which the world then abounded. Where are the monstrous lizards of the Wealden, the sight of whose mere skeletons makes us shudder? Where are the cave lions that once prowled through the woods of Great Britain, the cave tigers that hunted by their side, and the cave hyenas that munched the bones that have been found in European caves? As the world became prepared for higher beings, it became less fitted for them, and they perished. The crocodiles of to-day are the puny representatives of the huge saurians of the reptile age; and the diminutive lions and tigers of to-day, never seen in Europe, save harmlessly caged in the travelling caravan, are the degenerate kindred of the ferocious cave dwellers of the post-pliocene times. So, in the time to come, those that remain shall die; and in the language of Isaiah we may prophesy, "No lion shall be there, neither shall any ravenous beast go up thereon: it shall not be found there." "There shall be," so far as these are concerned, "nothing to hurt, and nothing to destroy."

Noxious insects must likewise disappear. Geologists know comparatively little about the insect inhabitants of the ancient world; but with a tropical climate everywhere, swampy lands, and rank vegetation, there must have been clouds of noxious insects, and of large size. From my own investigations, I know that our gnats and mosquitoes, our tarantulas and scorpions, are but a feeble few that represent the perished hosts of the tertiary and earlier times. Those that remain must in like manner perish; for the world is for man, and whatever wars with his highest well-being must disappear.

The land-surface of the earth will be greatly increased. At present, the land-surface of the globe is said to be only about three-elevenths of the whole; and the fish possess vastly more than the men. But there was a time in the early history of the earth, when water covered nearly, if not entirely, its whole face. During the Silurian period, we know that the land-surface was composed of but a few craggy islands, germs of the mighty continents that now exist. During the Devonian period, the islands were expanded, and many united. In the time of the coal-measures, continental areas were developed; but, even then, more than half of North America lay beneath the waves. The succeeding formations indicate longer and broader territory. Higher and higher rose the mountains, and the waters of the ocean retired.

Not only does the land-surface of the globe increase in consequence of its elevation, as I have shown in former lectures, but, as the earth cools, the water on its surface descends to greater depths, circulates through its crevices, and becomes a constituent of the rocks that compose its crusts, not excepting the hardest: some rocks contain even as much as twenty-five per cent. There was a time when the water of the globe was kept in the atmosphere by its intense heat; then a time when it rested upon its surface: and, even now, it does not descend more than two or three miles at the farthest; for at that depth it would be converted into steam. But, as ages pass away, the distance to which it can descend will increase; the superfluous waters of the globe will be drained off, and a much larger extent of land-surface be produced for man's occupancy.

It is a common opinion that the cooling process, by which volcanoes and earthquakes are to be destroyed, will produce such a diminution of the heat of the globe, that life will eventually be entirely destroyed in consequence of the extreme cold. It is true, there was a time when the temperature of the earth was extra tropical, even to the poles; but during the drift period, which probably lasted for hundreds of centuries, we know that the cold was so great in Europe, that Switzerland, Eastern France, and Northern Italy, were covered with a sheet of ice of enormous size and thickness, slowly moving over the land. The cold was so intense in North America, that a corresponding icy sheet covered it, extending from the Arctic Ocean as far south as Southern Massachusetts, Southern Ohio, and Northern Kansas; narrowing as it approached the western side of the continent, as the marks made by its motion, and still remaining, clearly testify. Gradually the domain of Winter has been diminished, his stern rigor relaxed; and wide realms have been delivered from his grasp. Even within the historical period, there seems to have been some climatal change in this direction. In the time of Strabo, the vine, it is said, could not be grown in Northern France; and the Rhine and the Danube were frequently frozen over. At the present time, the vine is not only grown all over France, but even to the north of it; and the Rhine and the Danube are very rarely frozen. It may be that the vine has become acclimated to a colder region since the time of Strabo, and that the accounts of ice on the Rhine and Danube are exaggerated. The amelioration of climate within the historical period may have been too slight to be recognized; but geology enables us to grasp such immense periods, that we can draw our conclusions with confidence. It is not so hot as it once was; it is not so cold: in both respects, the climate of a large part of the earth has improved. Even since the time of the stone men of the Valley of the Somme, when bowlders were floated down that river on ice, the climate of France has become much warmer; and we may fairly conclude, that, notwithstanding the decrease in the earth's internal heat, — which, if entirely absent, could not materially affect the external temperature, — the climate, by the operation of unknown agencies, is slowly becoming modified, and the world better adapted to the necessities of mankind.

If, however, these various improvements should not take place by the operation of natural forces, they may be effected by the instrumentality of man. For ages, the earth was modified by inorganic agencies; then unconsciously modified by the animals and plants living upon it: now it is being consciously modified by man, and will be much more so in the future. The earth was first fire-made, rude as the shapeless mass of iron drawn from a puddling furnace, and placed under the formative hammers; next water-made, the face of the earth washed, the fire-made rocks ground down, layer after layer of sediment deposited, and the metamorphic foundations securely laid. It was subsequently shell-made, plant-made, beast-made, and is now to be man-made, finished, and beautified. The savage, like the beasts that preceded him, knew not the work that was given him to do; but now we are learning the full significance of our mission in this direction, and our constantly-increasing intelligence is being used to bring the earth into the best possible condition to administer to our happiness.

If volcanoes do not die out and earthquakes cease by the operation of causes distinct from man's agency, they may yet be brought to an end by that agency, or their destructive force confined to such portions of the globe as he may determine. At one time, the lightning flashed at will, and no man in the presence of a thunder-storm could feel secure from the deadly bolt; and, a hundred years ago, he would have been a bold prophet who should have foretold that man would obtain the mastery over this destructive manifestation of electricity that we now possess, and bring down its vengeful fires on a slender rod in safety to the ground. The future shall do infinitely more for man than the past has done; for our ability increases with every step that we take, and that becomes easy to-day which seemed absolutely impossible yesterday. The more dangerous any force is that we have not mastered, the more useful it becomes when we have reduced it to obedience. Some method may be discovered by which the terribly destructive agent engendered in the earth's interior, and manifesting itself in volcanoes and earthquakes, may be transmitted quietly and safely into space after yielding its giant force for man's benefit. Marsh tells us, in his "Man and Nature," that "it is a very ancient belief that earthquakes are more destructive in districts where the crust of the earth is solid and homogeneous than where it is of a looser and more interrupted structure. Aristotle, Pliny the elder, and Seneca, believed that not only natural ravines and caves, but quarries, wells, and other human excavations, which break the continuity of the terrestrial strata, and facilitate the escape of elastic vapors, have a sensible influence in diminishing the violence and preventing the propagation of the earth-waves. In all countries subject to earthquakes, this opinion is still maintained; and it is asserted, that, both in ancient and in modern times, buildings protected by deep wells under or near them have suffered less from earthquakes than those the architects of which have neglected this precaution." Our ability to penetrate the earth's crust increases every day; and by numerous deep wells we may bring up harmlessly, and usefully employ, what is now the cause of such dire calamities. When earthquake-power drives our machinery, the holiday of the world will have come.

Nearly all active volcanoes are in the vicinity of the ocean: as Vesuvius, near the Bay of Naples; Ætna, on the Island of Sicily, and close to the Mediterranean. Of the two hundred and twenty-five active volcanoes, one hundred and fifty-five are situated on islands; and this suggests the method, as well as the possibility, of their extinction by man's agency. The sea has already extinguished thousands, — ay, millions; and the time may come when man shall apply this extinguisher to the still remaining fires, and volcanoes will be felt and feared no more. We have men in New England who would undertake to make a tunnel into the heart of Vesuvius, and let in the Mediterranean, and accomplish it too; and, however difficult such tasks may appear now, every year makes them easier.

If noxious weeds should still continue to grow, with no apparent natural diminution of their numbers, man will say to them, "Begone!" and they will disappear. When Australia was first discovered, there were no plants worthy of cultivation on the island, and no fruit that man could use to benefit except two or three sour berries. All the vegetable productions were weeds; that is, plants unserviceable to man. But intelligent labor has made a wonderful change. "Oranges, pine-apples, figs, bananas, grapes, mulberries, peaches, nectarines, alligator-pears, and guavas flourish side by side with wheat, corn, potatoes, and all the fruits, flowers, and vegetables of the temperate zone;" and the weeds have disappeared, that these might take their place.

"In China, with the exception of a few water-plants in the rice-grounds, it is sometimes impossible to find a single weed in an extensive district; and the late eminent agriculturist, Mr. Coke, is reported to have offered in vain a considerable reward for the detection of a weed in a large wheat-field on his estate in England."* The time will come when man shall possess the earth, and determine what plants shall grow upon it, as our best farmers and gardeners do on their lands; and, when that time comes, the day of thorns and thistles, and all troublesome weeds, will be forever over.

In like manner, destructive beasts, poisonous reptiles, and noxious insects, shall be extirpated by man, if they do not disappear by such a natural process of extinction as has swept away the megalosaurus of the oölite and the cave tiger of more recent times. The rifle has proved the master of the lion, and it rapidly disappears before the advance of civilization. In India, they are now confined to a few wild portions; and, in South Africa, they have vanished from all the fully-settled districts. Wolves and bears were quite common in Great Britain since the Christian era, but were extir-

* Man and Nature: Marsh.

pated in England in 1350; in Scotland, about 1600; and in Ireland, about 1700. The more numerous and intelligent the people, the sooner they obtain dominion over these foes. Lions and bears existed in Judæa in scriptural times. Panthers, wild-cats, rattlesnakes, and copperheads abounded in the Middle and many of the Northern States, where, at the present time, they are hardly ever seen. Where intelligent man goes, the bear retreats, growling, to the wilderness. The rattlesnake finds more danger in the glance of his eye than man does in its fangs; and, when man's dominion over the earth is complete, these destroyers of his peace will exist no longer.

Serpents are most numerous, and most poisonous, where man is least developed. Taking the general average, about one in five is poisonous; but, in Africa, one out of every three, and in Australia as many as seven out of ten, are poisonous. Intelligence has slain the most deadly. Who can doubt the extinction of the remaining, when the heirs of the world shall obtain their inheritance?

Mosquitoes and gnats are terrible pests of human kind, sucking out a man's blood and patience at the same time. In some localities in South America, the wretched inhabitants, to save themselves from the attacks of these pests, sleep with their bodies covered over with sand three or four inches deep; the head only being left out, and covered with a cloth. In some of the densely-wooded and swampy districts of Canada, I have seen the necks of the men and boys, working in the woods, swollen and bloody from the attacks of a small gnat, which renders a veil necessary for those who wish to avoid its painful bite. These troublesome insects are bred in stagnant water; and, when the water is made to run, they will run also: when a country is thoroughly opened and drained, their ravages cease. Neighborhoods once infested by them are now nearly free from their attacks; and those now cursed are but waiting for man's intelligent voice to bid the plague to cease.

Should the climate remain cold and inhospitable in winter, as at present, we shall then make an artificial climate better suited to our necessities, and more in harmony with our comfort. This, we may be told, is impossible: so once appeared what we have already accomplished.

There was a time when man never dreamed of warming himself, or preparing his food, by artificial heat. He pulled up the wild roots, picked the wild fruits, and swallowed the raw oysters and mussels as he wandered naked along the beach. A cave by the river-side, or a hollow tree, served him for a habitation; and here he crept when the winter storm blew, and lay like a wild beast in his lair till the warm breath of spring wakened him, gaunt and hungry, to life once more. Ages passed before he learned to make a fire and feed it; make a tent of skins, and warm it by a fire in the centre. By slow steps, he passed from rude tents, and ruder caverns in the ground, to stone huts, cabins, comfortable houses, and stately mansions, with a heating apparatus by which winter is shorn of his rigor, and a genial temperature surrounds us continually. By what possibility could a thousand North-American Indians have met together in council in the month of January, with the thermometer at zero? How easily we manage it! There is an iron box: we put into it coal from our Pennsylvania mines; take out the heat that it took in from the sunshine when it was growing, vegetable matter; diffuse it through the room, and snap our fingers at the frost; for we have made a climate to suit ourselves.

What we have learned to do for ourselves we shall some day learn to do for our plants, and render ourselves, in a great measure, independent of the exterior climate. Here and there the work has been commenced, on a small scale, in the shape of green and hot houses. These show us, and, still more, the Crystal Palace of Sydenham, and the Garden of Plants in Paris, how flowers can be made to bloom, birds to sing, and butterflies and bees give a summer-like appearance to grounds, while Winter holds in his stern embrace all surrounding lands. Why should the land, from which all our food is derived, either at first or second hand, lie idle for six months in the year, and, in some parts of the country, for eight? There is sand enough in New England alone to make glass enough to cover the whole of North America, from the fortieth parallel of latitude to the Polar Sea; there is iron-ore enough to make iron sufficient to support it, and thus enclose the whole land; at the same time, deficient as New England is in fuel, there is fuel enough to give a summer heat in the depth of winter to her entire domain.

But fuel, even now, is scarce and dear, and the moderate use of fires to warm our dwellings taxes our powers. How, then, shall we heat the whole land? Wood, of course, must make way for coal; and this, as its use increases, will diminish in price. The American coal-cellar is well supplied; and, for a long time to come, we shall draw from what some have designated "exhaustless resources." They are not, however, exhaustless. The coal of England is going at the fearful rate of eighty million tons a year, or six square miles of an average thickness of six feet; and already the question of fuel in the future has assumed a great importance in England. Our evil day will only be retarded; and, eventually, we must answer the same question. We are, however, now in a condition, or at least approaching a condition, that suggests an answer. Our oil discoveries have much diminished the importance of this question to us, drawing out of the earth, as we now do, a million gallons of petroleum a day. "But will not the oil give out?" Of course, in time; but that time is far distant. The oil-bearing rocks are of great thickness, and of vast extent. From the base of the Silurian formation to the top of the Devonian, is, we now know, veritable oil territory; the oil-bearing corals being found in all the limestones of these formations. As these rocks underlie fully one-half of the continent, the possible oil-ground is of immense extent. The gas-springs of Western New York indicate its existence there, though probably at great depth. New England may yet count it among her treasures; and, when we bore for it to depths of four or five thousand feet, we shall find deposits large and enduring, of which few at the present time dream. We shall burn it for fuel as well as for illumination; steamboats will cross the ocean by its aid, and locomotives run more swiftly and with greater ease and cheapness than before. This is a new servant that man has wakened from the sleep of ages; and we are but beginning to learn his powers, and their appplication to the supply of our needs. Nor does the amount of free, flowing oil, give us any idea of the immense amount of this material which the earth contains. Many limestones and sandstones are so saturated with oil, that it can be distilled from them with profit. Bituminous shales abound, from one ton of which from twenty to sixty gallons of oil may be distilled. I saw one bed of petroleum shale, partly in Utah, and partly in Colorado, that, on a moderate computation, contained forty thousand million barrels of oil. A bed of bituminous shale, thirty feet thick, underlies one-half of Tennessee, and contains much more oil even than this.

"But the oil will not last forever." True: we shall drain the land of oil as surely as England will exhaust her cellars of coal; and the question arises, What shall we do then? The earth is a great magazine of fire, and man will yet draw from it for his benefit. The Garden of Plants in Paris is heated by water from an artesian well eighteen hundred feet deep, which has a temperature of eighty-two degrees Fahrenheit, and is carried in pipes under the soil. A salad ground at Erfurt, in Saxony, heated by water from an artesian well, yields a profit of sixty thousand dollars a year to the proprietor. These are indications of the applications yet to be made of the interior heat of the earth. Shafts are sometimes sunk now to a depth of two thousand feet; and it would be possible to bore from the bottom of those to a farther depth of four thousand feet: at that depth, we should find a heat of at least one hundred and fifty degrees; and in some cases, by striking deep crevices, we should obtain a much greater heat than this.

Many hot springs are located on crevices which descend to such great depths that the high temperature of the earth's interior gives heat to their waters, and thus we may learn how to avail ourselves of this grand source of heat and power. For ages the savage knew not that the possibility of heat existed in the tree under whose shelter he lay.

By boring to a depth of a mile and a half (which is, perhaps, not impossible in the present condition of the mechanical arts, and may be easily accomplished in the future), we should have boiling water, supposing we had an artesian well, which we should probably have everywhere, at that depth. The steam that must exist there would be sufficient to drive up the water with great force; so that, in many cases, mechanical power could be derived from it as well as from the steam itself. The heated water from these wells could be carried into our houses, warming them most readily and economically. On a cold winter's morning we could turn on the hot water or steam, and supply ourselves with summer's heat without leaving our beds. It could be applied to cooking without difficulty, and supply the largest portion of the needs that wood and coal but inadequately and expensively supply at the present time. Laid under the soil, protected by glass, we should have a summer's heat supplied to our growing plants in winter, and bring the benefits of the tropics to our doors. Not only could we produce crops the whole year, but we might have in New England not only the apple, pear, plum, and peach, but the orange, fig, date, lemon, and pine-apple by their side.

I fix no time for these things to come to pass: it may be one thousand or ten thousand years. It has taken an immense period since man came into existence to advance to our present position, and much time will be necessary to take us where we are destined to go; but he who can judge of the tree by the sapling may know that these things shall be.

"But suppose, that, in time, the earth should become cold to a great depth, so that no more heat could be extracted from it by man: what then?" Even then we should not despair. Water, antagonistic as it is to fire, can, nevertheless, furnish us with the means to produce it. Water, as we know, is composed of oxygen and hydrogen gases, in the proportion of eight parts, by weight, of oxygen, to one of hydrogen; and these gases can be produced from water by several methods. If the ends of the platinum wires connected with a galvanic battery be placed near each other in a vessel of water, bubbles of hydrogen will rise from the one, and of oxygen from the other: the water is decomposed, and its constituent gases are thus produced. Hydrogen is a very inflammable gas, and oxygen is the supporter of combustion; and when united, as in the oxy-hydrogen blowpipe, they produce the most intense artificial heat. Platinum, which does not melt in the hottest furnace, fuses in the heat of this like wax. In water, then, we have stored up fuel for ages unnumbered. Our present methods of obtaining it are too expensive for ordinary use; but who can doubt that man's constantly-increasing ingenuity will invent some easy method of separating these valuable gases, and thus supplying us with needful fuel when other sources fail?

"But what shall we do," says a forward-looker, "when the water of the earth is all burned, the earth cold, the coal gone, and the oil consumed?" We shall never meet with such disaster; for this reason, — when hydrogen and oxygen gases, of which water is composed, are burned, water is produced by their combustion in exact proportion to the amount of the gases used; so that it might be burned over and over again forever. As long as the world exists, then, we may be assured that man's ingenuity will keep pace with his necessities of this kind, and the human race march on to the goal that shall lie before them.

Man is an important part of Nature; and his importance increases hourly. At first a helpless log, he floated on the stream, but now stems the current, or boldly directs it.

If the land-surface of the globe should not increase naturally in the future, as we have anticipated, man's agency would, without doubt, bring it to pass, as is evident from what he has already accomplished.

In Lincolnshire, England, four hundred thousand acres of fever-and-ague-breeding swamp-land have been transformed into fields of wheat, barley, and oats, and excellent meadows. In the Netherlands, lands lying still lower than the fens of Lincolnshire, and apparently much more hopelessly doomed, have been reclaimed, and become among the most productive. It has been calculated that nearly nine hundred thousand acres have been gained there by diking and draining. "The province of Zealand consists of islands washed by the sea on their western coasts, and separated by the many channels through which the Schelde and some other rivers find their way to the ocean. In the twelfth century, these islands were much smaller and more numerous than at present. They have been gradually enlarged, and, in several instances, at last connected by the extension of their system of dikes. Walcheren is formed of ten islets united into one. At the middle of the fifteenth century, Gœree and Overflakkee consisted of separate islands, containing altogether about ten thousand acres. By means of above sixty successive advances of the dikes, they have been brought to compose a single island, whose area is not less than sixty thousand acres." *

A few years ago, an English gentleman purchased for a trifling sum a small island which was covered by the sea every flood-tide, but left dry at the ebb. He enclosed it with a bank of earth thirty feet wide at the bottom, and seven feet high and four feet wide at the top, with a slope on the outside having two feet horizontal to one perpendicular. This wall, about two miles and a half long, encircled the island, except a gap about seventy feet wide, through which the tide flowed in and out. Earth was at first used to close the gap; but the sea swept it away as fast as it was thrown in. Piles were then driven in a double row, and clay rammed in between them. This succeeded, and the little island was drained. In time, excellent crops were raised upon it, a house and barn built, and nearly three hundred acres of land, by the energy of one man, won from the sea.

The draining of Lake Haarlem is one of the best examples that we possess of man's disposition and power to change water-surfaces into dry land; and is at the same time a prophecy of what will be done in the future, when the earth shall be as densely populated over its whole extent as it is now in Holland.

Here was a lake fifteen miles long, and seven broad in its greatest width. "What fine farms we might have here," said an enterprising Hollander, "if this lake were only drained!"—"Yes; but it lies below the sea-level, and it would be impossible to drain it."—"Then we must pump it dry."—"Pump it dry! Who ever heard of such an absurdity?" But pump it dry they did. For this purpose, three large steam-engines were employed, each

* Man and Nature: Marsh.

pumping a million tons of water in twenty-five and a half hours. They commenced pumping in May, 1848; and laid it dry in July, 1852. Where the boats sailed and the fishes swam are now comfortable cottages, fertile fields, and a population of five thousand thriving citizens. In the same country, it is now proposed to drain the Zuyder Zee, which covers two thousand square miles. The time will come when the land under Lake Erie will be of more value than the water within it; and, when that time comes, man will say to the waters, "March!" and they will go, leaving the land for man's occupancy. Its greatest depth is but two hundred and seventy feet, and its drainage would be an easy matter. In like manner, the lands of Lakes Michigan and Superior will be needed, demanded, and obtained, and the sea be made to give up a large portion of its shallow shores to supply man's constantly-increasing demand for room.

Many of these statements regarding the future are evidently based on the supposition that the population of the world will be much greater in the future. This will doubtless be the case. When the age of shells was, shells were so numerous that their accumulated remains made beds thousands of feet in thickness: universal ocean swarmed with testaceous inhabitants. When the age of plants was, so numerous were they, that the world's fuel for thousands of years to come is merely the excess of that vegetation then buried. Reptiles and beasts have in their turn held dominion over the earth, and their fossil remains indicate their existence in prodigious numbers, which had apparently for their habitation the universal earth. So, when the age of man shall have fully come, he will fill the world, and hold dominion over the entire globe. For one man, there will be at least a hundred men; and of course, for one woman, a hundred women; and, as a consequence, those who shall then live will enjoy existence much more than the solitaries of the present.

Every year finds the laws of health better known and more faithfully obeyed. We have learned that our diseases are, as a general thing, the consequences of our misconduct; and consequently, by avoiding the causes, we escape the consequences, and life is thereby lengthened. Hence, in nearly all civilized countries, the average duration of human life steadily increases. In Geneva, accurate registers have been kept for three hundred years. From 1560 to 1600, the average life of a citizen was twenty-one years two months; in the next century, twenty-five years nine months; in the century following, thirty-two years nine months; and in 1833, forty years five months. Thus, in less than three hundred years, the average duration of human life is nearly doubled among the constantly-increasing intelligent population of this Swiss city. In the fourteenth century, the rate of mortality in Paris was one in sixteen: it is now one in thirty-two. In England, the rate of mortality in 1690 was one in thirty-three; in 1780, it was one in forty. Yet, even now, millions die before their time. Multitudes of children are murdered before they see the light; others are born with enfeebled constitutions, owing to the ignorance of their parents: the same ignorance causes one-half of them to die before they are ten years of age. It is only by looking back a sufficient distance that we can see that any improvement has taken place; but, when we do thus look, the future is full of hope. The population of the United States increases with unexampled rapidity: the million and a half in New York and its suburbs — so young a city — is a fine example of this increase. The population of Canada increases at the rate of ten thousand a year. In London, there are six hundred more mouths to feed every week; and these are indications of the progress in population making in civilized countries generally. Thus, in time, will the whole earth be peopled by intelligent men; for this is evidently its destiny, toward which it has been striding from the earliest period of its history.

But how shall such a great population be fed? Already the crowded populations of Europe are making their way to our shores, and surging in wave after wave over our broad lands. There is no real necessity for this emigration from Europe. There are millions of acres of untilled land in Great Britain, that the poor would gladly cultivate if they were allowed. I have walked in England for thirty miles over one man's land, most of it in a state of nature, — poor land, it is true, but capable, by judicious culture, of producing excellent crops.

The lands that are cultivated will be much improved in the time to come. Our agricultural methods are still rude, and must be considered so while the plough is the symbol of agriculture. I can look back and see the first cultivator of the soil, a naked savage, who tries the first agricultural experiment by making a hole with his finger in the mud left by the overflowing waters of some river, and depositing a grain of wild corn. How he grins, on his return, to see the success of his experiment in the silken tassel, and eventually in the golden ears! He tries again; but, finding his forefinger sore from use, he improves on this by using a stick to dibble in the grains, and thus invents the first agricultural tool. His tribe learn from his example, and pursue the same course. But land suitable for their purpose becomes scarce, and the hard land must be used. A stone hatchet hews out the first rude spade, and with this wooden spade the work goes bravely on. I look down the stream of time, and see the wooden spade disappear, and a stick pulled by an ox take its place; follows the Saxon peasant with his wooden plough, *zeeing* over the ground, scattering his grain and reaping his scanty harvest, but thinking, proud soul! that never was such invention as his plough. What can possibly supersede it? Here comes the Yankee farmer with his steel plough and splendid team, turning over the broad furrow to the sun's eye, then with a harrow mellowing the rich soil, and with a drill sowing the grain in parallel rows: he fills his barns with plenty, and exclaims, "Where is the man that can improve upon this?" He is already born. I can see the great steam pulverizer, that shall supersede the plough as surely as the plough superseded the rough-hewn spade of the savage, stirring up the ground to the depth of two or three feet, and making it all as fine as the finest garden-mould. Then comes the steam-dibbler, planting every grain at an exact distance from every other; not one here, and another there, as at present, like the plums in a miser's pudding, or crowded together like the dollars in his bags, but springing up and standing like the chess-men on a board, with room enough to tiller out on every side. Weeds destroyed, the land will produce the crops, and nothing else, — crops that at this time would appear to us perfectly fabulous.

We shall not waste as we do now. Americans are doubtless the most wasteful people in the world. What is wasted in the United States alone would feed half of

Great Britain. We shall not use intoxicating drinks: they add nothing to man's bodily vigor, nothing to his mental ability. Think of the corn and potatoes wasted, worse than wasted, to make whiskey; grapes turned into wine and brandy, sugar into rum; to merely mention cherries, apples, peaches, transformed by infernal art into poisonous liquors. This must all cease, and the food wasted be redeemed for the service of mankind.

The tobacco used in the United States in a year, if taken at one dose, would destroy every human being on the globe. Hundreds of thousands of acres of the best land are set apart for the culture of this vile weed,—four hundred thousand in the United States alone, — which doctors, lawyers, parsons, and senators roll in their mouths, though as poisonous as the venom of the rattlesnake. The land and labor employed in its production shall likewise be redeemed; for, in the light of the day that is dawning, such demons as these shall vanish from the earth.

Land now considered worthless shall be made to yield abundance; for the desert must rejoice, and the wilderness blossom as the rose. Sahara is a name typical of barrenness and desolation; and yet see what science has accomplished even there! The French have bored a number of artesian wells in the desert near Algiers, and the natives are now doing the same thing. Beneath the sand lies a bed of clay; and, when this is perforated, up comes the water in a perennial fountain. On the first trial, after a few weeks' labor, a constant stream flowed out, yielding four thousand quarts of water a minute. Before the end of 1860, several tribes had abandoned their wandering life, planted palm-trees, and commenced the cultivation of the soil. Laurent, a French writer quoted by Marsh, says, "In the anticipation of our success at Oum-Thiour, every thing had been prepared to take advantage of this new source of wealth without a moment's delay. A division of the tribe of the Selmia, and their sheik, laid the foundation of a village as soon as the water flowed, and planted twelve hundred date-palms, renouncing their wandering life to attach themselves to the soil. In this arid spot, life had taken the place of solitude, and presented itself with its smiling images to the astonished traveller. Young girls were drawing water at the fountain; the flocks, the great dromedaries with their slow pace, the horses led by the halter, were moving to the watering-trough; the hounds and the falcons enlivened the group of party-colored tents; and living voices and animated movement had succeeded to silence and desolation."

Between 1856 and 1860, fifty artesian wells were bored, and thirty thousand palm-trees planted.

There is no land so poor that intelligence and industry cannot enrich it. The Chinese carry earth up the mountains, and deposit it on the bare rock, on which they succeed in raising valuable crops; and it is said that much of the wine of Moselle is obtained from grapes grown on earth carried up the cliffs on the shoulders of men.

The lands that man's ignorance and wickedness have cursed shall be blessed and redeemed by his intelligence and virtue. Virginia, doubly cursed by raising slaves and tobacco, shall become what it may easily be, — the orchard of America. Palestine shall yet be what it is said to have been, "a land flowing with milk and honey." The Valleys of the Tigris and Euphrates shall be regenerated and repeopled by a greater population than

Nineveh or Babylon ever knew. With free governments that will guarantee men the fruit of their labor; above all, with communities, which must be eventually established, laboring for the general welfare, and aided by all the capital and intelligence of a neighborhood combined, — the run-down, worn-out old lands of Asia and Europe shall be made once more to yield their increase, and administer to man's well-being.

The earth was all a desert once; and, after so much of it has been made fertile without our agency, it would be pitiful if we could not finish the remainder. The time and labor and life spent by the world in war would have pounded every bowlder in the land to dust to make soil, bored a hundred thousand artesian wells, and drained every swamp; it would have made — used in accordance with our highest modern culture — every desert a garden, and every wilderness a fruitful field. War is not to last forever: it is the game of fools; and, when men are wise, they will play it no more. The energy, the wealth, the increased intelligence, of mankind, will be employed to overcome the physical evils that surround us; and then the millennium that man is commissioned to make will have come.

The question is sometimes asked, "Will a new race of beings ever inhabit the earth as much superior to the present as they are superior to the forms that preceded them? Will the time ever come when man will be superseded by some superior being, and he be only known by fossil remains as an extinct form, — one of the shapes assumed by life in its march though the ages, and laid down when its purpose was answered, and a more fitting one assumed?" If the reptiles of the Carboniferous could have reasoned from the then past history of the globe, they would have doubtless said, "The world was made for us: what intelligent reptile can doubt it? Life advanced, during long ages, from the radiate to the mollusk and articulate, from them to the fish, and lastly to us, its highest representatives. Behold the perfection of our forms! See these lungs that breathe the vital air, which all others have obtained from the water by clumsy gills; these limbs that elevate us above the ground, and by which we can swim in the water, or walk or run upon the land! What reptile can conceive of a more perfect form than ours? Nature exhausted her resources in its production. The world was made for reptiles, and will be a reptile world as long as it endures."

When birds made their appearance, how could they doubt that they were creation's fairest, best, and last ordained? "What being can possibly surpass us? We are at home in the water, on the land, and, superior to all, in the air. It is impossible that any being should ever arise superior to us." May we not, in like manner, be flattering ourselves, looking only at the past, and not at the possibilities of the great future?

I regard man as the fruit of the tree of life; and, if he is, beyond the fruit the tree cannot go. A tree advances from root to stem, from stem to branch, from branch to leaf, and from leaf to blossom and fruit, each rising in importance above the other; but, when the fruit is attained, all that can be done is to perfect it. The root of the great tree of life is the radiata, their raying, ramifying arms and fingers forming its spreading radicles; the trunk of this tree, the mollusca; their shelly covering, its bark. The jointed bodies of the articulates form its branches: the vertebrates are the leaves. Every leaf has a mid-rib passing through its centre, from which ribs go to each side of the leaf to strengthen it, as in vertebrates the back-bone passes through the centre of the animal, and ribs proceed from it on both sides. The blossoms are the mammalia or milk-producing animals; and its fruit humanity, waiting for the ages to ripen it. This grand old tree has been advancing for ages, renewing its rootlets, shedding its old bark, losing unnumbered branches in the storms of the past, and dropping myriads of leaves and blossoms, but, with a sound heart, reproducing better than it lost, and fruiting in good time with the promise of the best when that fruit is fully ripe. But what evidence is there that man is the fruit of this wonderful tree? What peculiarity is there in the fruit of a tree that distinguishes it from every other part? It contains a living principle which possesses unlimited duration, and, under favorable circumstances, may unfold into a tree equal or superior to that from which it sprang. Let a piece of the root be separated from the tree, it speedily dies, and is resolved to dust: in like manner, bark, branches, blossoms, leaves, perish when their connection with the parent plant is dissevered. The fruit alone contains the power of continuous existence within itself. Drop it on the ground, or bury it, and it lives and grows, and sends its type down the ages: so man. The polyp, the snail, the worm, the fish, reptile, bird, and beast may die when death comes, and return to the undistinguished dust from which they sprang; but man possesses that over which death has no power, and the extinction of one life is but the dawn of another of greater power and beauty. Some there are who doubt this; to such this argument will have no weight: but to those who believe in the soul's future, and to others, who, like myself, know that we continue to live hereafter, the reasonableness of this will be apparent.

Did not man possess the power of unlimited progress, he would be dropped for some form superior to him in this respect. Nature progressed in the fish till the fish could advance no farther and be a fish; she then progressed in the reptile, till, in the pterodactyle and allied forms, they could advance no farther and be reptiles; she then chose the bird, and for the same reason left it behind, and took up the beast; she has now chosen man in whom to embody this principle, and in him she finds that power of unlimited progress which satisfies her. We stand on the platform that our progenitors built, and we build higher for our children: in their turn, they elevate it for the next generation. As a race, then, we satisfy the law, and as individuals. The great future opens its portals for us, and presents us a boundless field for our advancement.

As the earth is being gradually cured of its evils, and as its organic forms have been manifested in continually progressive forms, so we may reasonably expect a superior race of human beings, and the eventual destruction, by the growth of the superior faculties, of the moral evils that war with our highest interests. As we have outgrown cannibalism, to which our forefathers were addicted; as we have advanced from the wild savages with rude stone weapons that hunted the mammoth through the woods of Great Britain, and dwelt in caves by the shore: so shall we outgrow war, intemperance, licentiousness, lying, bigotry, and every form of wrong-doing, and grow into intelligence, culture, and every manly virtue, — lovers and loved of all.

What will be the final destiny of the earth? As there was a time when the world was not, so there will come a time when it will cease to exist. When fruit-trees can produce fruit no longer, they die, and return to the earth, to give place to those that can produce fruit in turn; and when the earth is old, worn out, and can no longer administer to man, then we may reasonably expect that it will die, and return to the sun, from which it probably came. It has been a common opinion that the heavenly bodies revolve in a vacuum: but scientific men now generally believe that space is occupied by some resisting medium; and, however finely attenuated it may be, its influence upon the planets moving in it must be such as to retard their motion, and eventually bring them to the sun. Littrow, as quoted by Mayer, says, "The assumption that the planets and comets move in an absolute vacuum can in no way be admitted. Even if the space between celestial bodies contained no other matter than that necessary for the existence of light (whether light be considered as emission of matter or the undulation of a universal ether), this alone is sufficient to alter the motion of the planets in the course of time, and the arrangement of the whole system itself. The fall of all the planets and the comets into the sun, and the destruction of the present state of the solar system, must be the final result of this action."

The smaller any cosmical body is, the faster it will move toward the sun: hence, although we cannot observe this tendency in the large planets, on account of their size, in some of the comets it is plainly visible. Encke found that the comet named after him, which revolves around the sun in 1,207 days, has its motion so accelerated as it approaches the sun, that the time of such revolution is shortened by about six hours. The comets are grand celestial moths flying round and round the central luminary, and eventually destined to plunge into it and end their dazzling career. In their fate we may see our own. One by one, the planets shall be gathered home. Mercury and Venus shall march to their graves before us, and perhaps enable us to calculate the time of our endurance. The sun itself may eventually return to the grand central sun around which it revolves, and the matter composing it be again sent out in suns and planets in the great eternity of the future, as may have been the case in the equally great eternity of the past.

GLOSSARY

OF

SCIENTIFIC AND DIFFICULT TERMS USED IN THIS VOLUME.

Those defined in the Text may be found in the Index.

ADAMANTINE, like adamant; which is a name given to the diamond and other bodies of great hardness.

AGGLOMERATED, collected into a ball.

AMPHIBIAN, an animal so constituted as to be able to live in water and on land.

AMYGDALOIDAL TRAP (*amygdala*, an almond), one of the forms of trap-rock in which enclosed minerals are embedded in roundish bodies.

ANTHRACITE, a bright, hard coal that burns without smoke or flame, the bitumen having been driven off by heat.

ARCHEOPTERYX (*arche*, beginning; *pteryx*, a wing), a reptilian bird of Solenhofen.

ARID (*aridus*, dry), parched with heat, destitute of moisture.

ARMADILLO, an insectivorous mammal of South America. covered with a hard, bony shell composed of movable bands.

ARMATURE, armor; that which defends the body.

BITUMINOUS, containing bitumen, an inflammable substance, found of varying consistence, from naphtha to asphaltum.

BOWLDER, a rounded mass of rock, generally found at a distance from its original position.

BRECCIA, a rock made up of angular fragments cemented together.

BRYOZOAN (*bryon*, moss, and *zoon*, an animal), small mollusks which grow together on a common stock, and so much resemble polyps that they were formerly arranged with them.

BUCKLER, the shield or body-covering of an animal.

BURIN, a sharp-pointed instrument used by engravers for marking.

CANADA BALSAM, a kind of turpentine obtained from the balm-of-Gilead fir.

CANNEL-COAL, originally candle-coal, and so called because it is an inflammable coal that will burn like a candle.

CANNELITE, a highly bituminous shale, generally found in the neighborhood of petroleum coal.
CANYON, a chasm cut by a river with walls nearly or quite perpendicular.
CARBONIC-ACID GAS (*carbo*, coal), a heavy, poisonous gas composed of one part of carbon and two parts of oxygen. It frequently issues from the ground in volcanic countries, and is often found in wells and mines.
CARBURETTED-HYDROGEN GAS, an inflammable gas composed of carbon and hydrogen, which is very abundant in coal-mines.
CARICATURE, an exaggerated figure, bearing only a distant resemblance to the object.
CARPUS, that part of the skeleton which forms the wrist.
CHALCEDONY, a translucent variety of quartz.
CHEIROMANCER, one who tells fortunes by the hand.
CHERT, an impure variety of flint.
CHIMPANZEE, an ape found on the western coast of Africa, which, next to the gorilla, most resembles man.
CLIMATAL, relating to climate.
CLUB-MOSS, a moss-like plant intermediate between a fern and a moss.
CONCENTRIC, having a common centre.
CONCHOIDAL (*concha*, a shell), having curved elevations and depressions in form, like the valve of a shell.
CONGLOMERATE, rock made of pebbles cemented together; sometimes called pudding-stone.
CONICAL, in the shape of a cone or sugar-loaf.
CONIFER, a plant bearing cones. The pines and firs are conifers.
CONIFEROUS, bearing cones.
COPROLITE, fossil dung.
CRANIUM, the skull of an animal.
CRATER (*crater*, a cup), the circular cavity of a volcano, from which the matter of an eruption is ejected.
CRESSET, a great light set on a watch-tower.
CRUSTACEAN, an articulate animal covered with a crust or shell, like a crab or lobster.
CYCAD, a plant growing in warm countries, which is intermediate in its structure between pines and palms.
CYSTIDEAN, a radiate animal belonging to the family of crinoids. They are only found in a fossil state.

DÉBRIS (pronounced dā-bree′), fragments detached from rocks, and piled up in masses.
DETRITUS, sand, fine gravel, or clay, worn from solid bodies by water or ice.
DIGIT (*digitus*, a finger), a finger.
DILUVIUM, in this volume it means the time when the glacial beds were deposited.
DISINTEGRATING, wearing down to fine particles.

ECHINITE, a fossil echinus.
ECHINODERM (*echinus*, hedge-hog; *derma*, skin), a radiate animal, generally of a nearly globular form, and covered with spines. The sea-egg is a familiar example.
ECHINUS, same as echinoderm.
EDIBLE, eatable.

ELEMENTS, those bodies which cannot be decomposed by the chemist, and of which all other bodies are composed.

ENAMEL, the hardest portion of a tooth; the smooth, hard substance which covers or composes the most elevated parts of the tooth.

ENTOMOLOGICAL, relating to the science of insects.

EPHEMERA (*ephemeros*, daily), a fly that lives but for a short time; strictly, a fly that lives but for a day.

ESTUARY, the place where a river pours into the sea or a lake.

FAHRENHEIT THERMOMETER, a thermometer invented by Fahrenheit, in which the freezing-point of water is marked 32°, and the boiling-point 212°.

FATHOM, six feet.

FILIGREE-WORK, an ornamentation on gold and silver, worked in the manner of little threads or grains.

FLUOR-SPAR, a mineral composed of the two elements, fluorine and calcium. Its crystals are of beautiful colors.

FRITH, a narrow passage of the sea, or an opening of a river into the sea.

FUCOID, sea-weed.

FUCOIDAL, of a fucoid.

GORILLA, of apes the most man-like. It is found on the western coast of Africa.

GREENSTONE, a variety of trap of a green color, containing hornblende and felspar in small crystals.

GUACHOS, the men who live on the South-American pampas.

HOMOGENEOUS, of the same kind or nature throughout.

HOMOPTERA, an order of insects having four membranous wings. They feed on the juices of plants.

HORSE-TAIL, a plant that grows in boggy places, and allied to the ferns.

HUMERUS (*humerus*, shoulder), the bone of the arm nearest the shoulder.

HYDRAULIC CEMENT, a cement composed of lime and clay that sets under water.

HYDRO-CHLORIC ACID (sometimes called muriatic acid), a very powerful acid composed of hydrogen and chlorine.

ICHNOLOGICAL, relating to the science of fossil footprints.

IGNEOUS, fiery.

INDIGENOUS, originally belonging to a country.

INFILTRATION, the act of entering the pores of bodies.

INFUSORIA, animals invisible to the naked eye, and often found in vegetable infusions.

INTEGUMENT, that which naturally surrounds a thing.

INVERTEBRATE, destitute of a back-bone. Radiates, shells, and insects are invertebrates.

KING-CRAB, an articulate animal living on the coast of New England; sometimes called horse-foot. Its scientific name is *limulus*.

LAGOON, a shallow pond or lake into which the sea flows.

LARVÆ, insects in the caterpillar or grub state.

LAVA, the melted matter that flows from a volcano.

LENTICULAR, having the form of a lens.

LIGNITE, wood partially converted into coal.
LIVERWORTS, an order of plants having short, leafy stems, and allied to the mosses.

MAGNESIAN LIMESTONE, limestone containing magnesia.
MARINE, belonging to the sea.
MATRIX, a mould; the cavity in which a thing is held.
METACARPUS, the part of the hand between the wrist and fingers.
MOSS-AGATE, a chalcedony containing crystallized manganese, which has a plant-like appearance.

NEBULÆ, cloudy spots in the heavens, some of which are assemblages of stars, and others gaseous matter uncondensed.
NEBULOUS, having the character of nebulæ.
NECROMANCER, one who tells fortunes by departed spirits.
NON-NITROGENOUS, containing no nitrogen.

OCCIPITAL BONE, the bone at the back part of the head.
ORGANIC, that which possesses organs by which it grows to perfection, as plants and animals.
ORNITHORYNCHUS, a singular mammal of New Holland, which possesses a bill and webbed feet like a duck.

PAMPAS, the plains of South America which are destitute of trees.
PENTAGON, a figure hav ing five corners.
PERCOLATED, passed through the interstices or small holes of.
PHALANGES, the small bones of the fingers and toes.
PLACO-GANOID, partaking of the nature of the placoidal and ganoidal fishes.
PLASTIC, capable of being moulded.
POLYPS, radiate animals surrounded by arms or tentacles, which live in communities, and form coral by the secretion of limy matter.
POST-PLIOCENE, the period indicated by beds, all of whose fossils are identical with living species.
PRAWN, a small crustaceous animal of the shrimp family.
PRIMORDIAL ZONE refers to that class of rocks which are supposed to contain the earliest forms of life; generally applied to certain beds of Bohemia.

RADIUS, distance from the centre.

SALAMANDER, a frog-like reptile, whose organization seems to be between the frog and the lizard.
SALIFEROUS, salt-bearing.
SALINE, salty.
SCAPULA, the shoulder-blade.
SELENITE, crystallized sulphate of lime.
SERRATED, resembling a saw.
SHALES, fine-grained rocks, generally softer than slates, but in other respects resembling them.
SHEIK, an Arab chief.
SIDEREAL, belonging to the stars.
SILICIFIED, petrified by silica or flinty matter.

SIMOOM, a hot, dry wind that blows occasionally in Africa and South-western Asia.

SPECTRUM ANALYSIS, a method of telling the character of light by passing it through a glass prism, and observing the dark lines that cross the spectrum.

SPHERICAL, having the shape of a sphere or ball.

STEPPES, the Russian name for the large plains of Northern Asia.

STIGMARIA (*stigma*, a mark or impression), a genus of plants found in the coal-measures. No plants at the present day resemble them. They derive their name from the marks left on the trunk by the leaves when they dropped.

STRATUM (plural, strata), a bed or layer.

SUBMERGENCE, the state of being under water.

SUTURE, the seam which unites the bones of the skull or parts of a shell.

TABULATED, having a flat surface.

TABULÆ, plates having a flat surface.

TALCOSE-SCHIST, a slaty rock containing talc.

TARANTULA, a stinging-spider.

TENTACLES, thread-like processes, generally surrounding the mouths of animals, by which they feel, and convey particles of food to the mouth.

TEREBRATULA (*terebro*, to bore), a family of mollusks in which one of the valves has a hole for a ligament, by which the animal attaches itself to bodies.

TESTACEOUS, possessing a shell, as a mussel or oyster.

THRIPS, an order of insects having long membranous wings. They commit great depredations on plants.

TRANSITIONAL FORM, an organic form intermediate between two distinct kinds.

TRAP, ancient lava. A heavy rock of a greenish color. The word is derived from *trappa*, a Swedish word for "stair," because rocks of this kind are frequently found in masses, one above another, like the steps of a stair.

TRAP-DIKE, a crevice filled with trap.

TREE-FERN, a genus of ferns very common in the coal-measures, and which are now found in Australia.

TREMOR, a trembling.

TRIGONIA (*treis*, three; *gonia*, angle), a tribe of mollusks possessing shells having three corners.

ULNA, the larger of the two bones of the fore-arm

URUS, the wild bull.

VACUUM, empty space.

VERTICAL, directly overhead.

ZOÖPHYTE (*zoon*, animal; *phytus*, plant), a term applied to radiate animals that resemble plants.

INDEX.

www.ingramcontent.com/pod-product-compliance
Lightning Source LLC
LaVergne TN
LVHW010207110826
845151LV00002B/631

9781425535438